AF901297

ROUTLEDGE LIBRARY EDITIONS:
THE AUTOMOBILE INDUSTRY

Volume 3

TECHNOLOGICAL CHANGE, RATIONALISATION AND INDUSTRIAL RELATIONS

TECHNOLOGICAL CHANGE, RATIONALISATION AND INDUSTRIAL RELATIONS

Edited by
OTTO JACOBI, BOB JESSOP,
HANS KASTENDIEK AND MARINO REGINI

Routledge
Taylor & Francis Group
LONDON AND NEW YORK

First published in 1986 by Croom Helm

This edition first published in 2018
by Routledge
2 Park Square, Milton Park, Abingdon, Oxon OX14 4RN

and by Routledge
605 Third Avenue, New York, NY 10017

Routledge is an imprint of the Taylor & Francis Group, an informa business

© 1986 Otto Jacobi, Bob Jessop, Hans Kastendiek and Marino Regini

All rights reserved. No part of this book may be reprinted or reproduced or utilised in any form or by any electronic, mechanical, or other means, now known or hereafter invented, including photocopying and recording, or in any information storage or retrieval system, without permission in writing from the publishers.

Trademark notice: Product or corporate names may be trademarks or registered trademarks, and are used only for identification and explanation without intent to infringe.

British Library Cataloguing in Publication Data
A catalogue record for this book is available from the British Library

ISBN: 978-1-138-73855-3 (Set)
ISBN: 978-1-315-16182-2 (Set) (ebk)
ISBN: 978-1-138-03880-6 (Volume 3) (hbk)
ISBN: 978-1-315-17710-6 (Volume 3) (ebk)

Publisher's Note
The publisher has gone to great lengths to ensure the quality of this reprint but points out that some imperfections in the original copies may be apparent.

Disclaimer
The publisher has made every effort to trace copyright holders and would welcome correspondence from those they have been unable to trace.

TECHNOLOGICAL CHANGE, RATIONALISATION AND INDUSTRIAL RELATIONS

Edited by

Otto Jacobi, Bob Jessop, Hans Kastendiek and Marino Regini

CROOM HELM
London & Sydney

©1986 Otto Jacobi, Bob Jessop, Hans Kastendiek and Marino Regini
Croom Helm Ltd, Provident House, Burrell Row,
Beckenham, Kent BR3 1AT

Croom Helm Australia Pty Ltd, Suite 4, 6th Floor,
64-76 Kippax Street, Surry Hills, NSW 2010, Australia

British Library Cataloguing in Publication Data

Technological change, rationalisation and industrial
relations.
1. Technological innovations – Great Britain
2. Industrial relations – Great Britain
I. Jacobi, Otto
338'.06 HC260.T4

ISBN 0-7099-1448-2

Printed and bound in Great Britain by Mackays of Chatham Ltd, Kent

CONTENTS

Preface
Abbreviations
Introduction
Between Erosion and Transformation: Industrial Relations Systems Under the Impact of Technological Change. Otto Jacobi, Hans Kastendiek, and Bob Jessop. 1

PART ONE TECHNOLOGICAL CHANGE AND LABOUR RELATIONS 13

1. Technological Innovation, Organisation of Work, and Unions. Federico Butera and Giuseppe Della Rocca........................ 15
2. Changing Skill Requirements and Trade Union Bargaining. Zissis Papadimitriou........... 35
3. Technological Change, Labour Market, and Trade Union Policy. Gerhard Brandt......... 50

PART TWO THE POLITICS OF RATIONALISATION: THE CAR INDUSTRY 71

4. Rationalisation and Industrial Relations: a Case Study of Volkswagen. Eva Brumlop and Ulrich Juergens........................... 73
5. The Politics of Technological Change at British Leyland. Harry Scarbrough.......... 95
6. Changes of Industrial Relations at Fiat. Matteo Rollier............................ 116

PART THREE CRISIS AND RATIONALISATION: IMPACT ON UNIONS 135

7. Bureaucracy, Oligarchy, and Incorporation in Shop Steward Organisations in the 1980s. Eric Batstone.............................. 137
8. Shop Stewards and Management: Collective Bargaining as Co-operation. Michael Terry. 161
9. Some Current Strategy Problems of the Italian Trade Unions. Hartwig Heine........ 176
10. Centralisation or Decentralisation? An Analysis of Organisational Changes in the Italian Union Movement at a Time of Crisis. Ida Regalia................................ 196
11. Social Change and Trade Union Movement in the 1970s. Aris Accornero.................. 219
12. Labour Conflicts and Class Struggles. Walther Mueller-Jentsch.................... 238
13. Workers' Reactions to Crisis. Rainer Zoll and Enno Neumann.......................... 261

Notes on Editors and Contributors 276

List of Translators 279

Index 280

PREFACE

This volume, like a parallel one on "Economic Crisis, Trade Unions and the State", is a result of a tri-national co-operation. Most of the contributions by British, Italian and West German colleagues were originally presented to two conferences at Cambridge and Torino in autumn 1983 which we organised in the course of a research project at the Institut fuer Sozialforschung, Frankfurt am Main. The "production" of both readers, i.e., the preparation and subsequent revision of the contributions as well as the editing, was itself a matter of concrete co-operation. We thank all those who took part in the whole "enterprise", especially Marino Regini who very effectively acted as co-editor for the Italian contributions and Ingrid Zierold who works with us in Frankfurt. Bob Jessop performed a key role in co-ordinating the editing and preparing the text for publication. We hope that both volumes prove our joint efforts to come to ends with some of the major problems discussed in the three countries.

The present volume demonstrates that comparative studies have to refer to similar or identical problems in the countries concerned but also have to take into account quite specific problem constellations. Thus the Introduction and Part 1 outline some of the general problems of "Technological Change and Labour Relations". Part 2 aims at a sectional comparison by discussing the politics of rationalisation and of industrial relations in the car industry. The "cases" of Volkswagen, British Leyland and FIAT are both: an exemplification of the general theme and a specification of national adjustments to economic crisis, technological change and rationalisation. Part 3 has a slightly different outline. Some of the impacts present developments exert on trade unions can be identified to a same extent in all of the three countries and therefore are discussed simultaneously. Others get their weight from specific national conditions and constellations and therefore demanded specified consideration. What for Part 3 may be seen as a topical asymmetry is in fact a result of typical differences which we did not want to cover by a schematic presentation.

This argument, of course, does not relieve comparative research of further efforts to cope with substantial, theoretical and methodological problems of cross-national studies. Studies like these, however, depend even more than others on adequate resources. In our case, the Volkswagen Foundation granted the Frankfurt project mentioned above, and additional subsidies were given by the German Research Foundation and the Goethe-Institut at Torino.

June 1984

Otto Jacobi Hans Kastendiek
Institut fuer Sozialforschung
Frankfurt am Main.

ABBREVIATIONS

ACAS	Arbitration, Conciliation, and Advisory Service
BIW	Body in White
BL	British Leyland
BLMC	British Leyland Motor Company
CAD	Computer-aided Design
CAM	Computer-aided Manufacturing
CAP	Computer-aided Programming
CIR	Commission on Industrial Relations
CNC	Computer-aided Numerically Controlled
CO2	Carbon Dioxide
CPU	Central Processing Unit
CGIL	Italian General Confederation of Labour
CISL	Italian Confederation of Labour Unions
DNC	Directly Numerically Controlled
EDP	Electronic Data Processing
EEC	European Economic Community
FRG	Federal Republic of Germany
IG Metall	Industriegewerkschaft Metall (Metal Industry Trade Union)
ILO	International Labour Organisation
LODI	Lohndifferenzierung (wage differentiation)
MDW	Measured Day Work
NC	Numerically Controlled
NBPI	National Board on Prices and Incomes
R and D	Research and Development
T&GWU	Transport and General Workers' Union
UAW	United Automobile Workers
UIL	Italian Union of Labour

INTRODUCTION

BETWEEN EROSION AND TRANSFORMATION: INDUSTRIAL RELATIONS SYSTEMS UNDER THE IMPACT OF TECHNOLOGICAL CHANGE

Otto Jacobi, Hans Kastendiek, Bob Jessop

1. Technological change and rationalisation of production and/or the labour process have always accompanied periods of economic, social, and political development in capitalist societies. Crises have frequently been triggered or exacerbated by the obsolescence of long-established production and labour processes and its inhibitory effects on economic growth and capital accumulation. Conversely, recovery from a major economic crisis has often been accompanied by the introduction of a new technology, which creates new markets and facilitates increased productivity. Throughout its history, industrial development has seen cycles in which a phase of renewal brought about by dramatic technical advance is followed by a complementary phase of maturation in which the existing technical and social aspects of the labour process are improved and refined. In the period of innovation, particularly at plant and sectoral level but also in the economy as a whole, technological change and rationalisation have led to abrupt adaptation. Conversely, the longer periods of diffusion and maturation which follow have typically involved more gradual processes of modification.

Technological development has always been crucial for the dynamic of the relations between labour and capital and between the organisations which represent them. The development of industrial relations systems does not, however, appear to be fully determined by technological change. Indeed, there are only limited chronological parallels with its successive phases of innovation and maturation. Industrial relations systems as political, economic, and social phenomena are influenced by historical and cultural factors. Thus they tend to be more closely connected with the long-term economic situation and the more general balance of social and political forces than they are with short-term, cyclical fluctuations in the market. This implies a considerable degree of inertia in industrial relations systems and these do, indeed, differ widely across countries with similar technological bases. But one cannot ignore the effect of technological change on industrial relations in the longer term. The traditional structure of the labour force has been thrown into disarray with the demise and emergence of a variety of trades, the disappearance of traditional professions and qualifications as they are replaced by new ones, and the Taylorisation of labour. In response to new challenges, labour and trade union movements have adapted internally through continuing processes of education, politicisation, and mobilisation of the labour forces; and have reinforced their bargaining power externally through new forms and practices of solidarity and collective representation.

There is currently an obvious tendency towards industrial or craft unions which are not divided along political, religious, or other non-economic lines in all countries. They have established themselves as bargaining forces in relation to capital in the economic sphere and in relation to the state in the political realm. But this tendency cannot be explained satisfactorily without referring to the equalising effect of the adoption of new technology by industry. All advanced industrial societies now have access to essentially the same technology. Inevitably this results in some convergence among industrial relations structures. In part this erodes the historically and culturally coloured national character of the system in a particular country and in part it is absorbed by this national specificity in distinctive ways. If convergent tendencies result from the equalising effect of applied technology on social structure and labour organisation, then divergent responses and developmental patterns must be due to the socially constituted and mediated development and application of technology. Used or desired as an instrument of domination, therefore, technology will always be the subject of conflicts. How these conflicts are resolved will vary over time according to conjunctural changes in the balance of forces and fundamental struggles for power as well as across societies according to the historically- and culturally-based ideologies and objectives of the actors involved.

Introduction

2. Technological development in the first three decades of the postwar era was linked to a phase of expansion and maturation. It was characterised by a consistently broadening range of applications and the international diffusion of existing technology. This provides an excellent opportunity to observe the flexibility and inertia of industrial relations in the context of economic development and technological change. Comparison of some fundamental features of development in Great Britain, the Federal Republic of Germany, and Italy shows the following picture:

- In Great Britain, a low growth rate, a relatively retarded level of technological advance, and a largely unchanged system of industrial relations which had no political or social steering function, all fused into the vicious circle of an imperialist power in a protracted period of decline. The comparatively low rates of economic growth and technological progress show how ill-equipped the country was to adjust to the prevailing political and economic situation. Attempts to compensate for the loss of guaranteed markets within the Empire by forward thinking and adaptation to competitive international markets were largely unsuccessful. The inability to innovate - a failing which cannot be laid at the door of the unions alone - is also evident in industrial relations. The relatively under-institutionalised and only minimally legalised system of regulations, the fragmented collective bargaining system, the multiplicity of labour organisations, and the low level of centralisation - all inherited from the turn of the century or the period between the wars - proved extraordinarily resistant to attempts at change. Neither the wartime coalition government nor the many subsequent state-initiatives provided a lasting impetus for structural innovation. Since the historical continuum was less severely disrupted by the war in Great Britain than in Germany or Italy, inertia outweighed the capacity for renewal. The comparatively high level of social conflict was a result of the mutually reinforcing interaction of economic and technological retardation and the inherited defects of the political and social steering systems. Both Germany and Italy were forced by the inescapable facts of war and fascism to make a new start. A voluntary initiative was required, however, if Britain was to adapt to political and economic changes. At first, the need for change was ignored and, when the attempt was finally made, it was not exactly successful.

- The postwar years in the Federal Republic of Germany, on the other hand, saw economic growth and expansion, rapid technological development, and dynamic industrial relations. These factors merged into what might be called a "virtuous circle" for a country which was staking its very survival on the new start. The strength of the West German model lies in the orientation of the national production apparatus towards international markets and in the determination to make good technological deficits with other advanced industrial economies. Together these secured a competitive advantage via technological superiority. With continuing economic expansion and political consolidation, the labour movement gradually shrugged off all remnants of class struggle militancy and allied itself with the traditions of reformism and economic democracy.

The export- and modernisation-orientated growth model was accepted by the trade unions. They believed it would produce high levels of employment and productivity. In turn this would strengthen their bargaining power and create a sufficiently large economic surplus to raise the individual and collective living standard of the labour force. The tripartite consensus produced a growth alliance which was greatly reinforced by the development of negotiating structures which were functional for system maintenance. The high level of centralisation, legal recognition and institutionalisation of the West German industrial relations system are a legacy of German social history. Government legislation, together with case law and management-workforce agreements gradually moulded the industrial relations system to the requirements of a national economy dependent on world markets and committed to technological progress. The dual structure of collective representation, the limited and equally (paritaetisch) distributed means of struggle, and the concentration of internal and external authority at the top of the unions all contributed to a flexible approach to problem-solving, the ability to absorb conflicts, and a readiness to compromise. The manner in which agreements between management and workforce applied at plant, trade, and national levels, combined with the fact that the workforce shared in the resulting wealth, provided West German capitalism with the necessary degree of effectiveness and legitimacy. Thus the comparatively low incidence of conflict results not only from the

German labour movement's tradition of social partnership or its willingness to be bribed through the dynamic of market forces, but also derives from institutionalisation of the class struggle.

- Of the three countries under consideration, the most enduring changes have occurred in Italy. Almost explosive economic growth, industrialisation and mechanisation fundamentally altered the structure of industrial relations. The process was beset by conflict, and produced the type of "virulent" circle which is typical of late developers. At the end of the war, Italy was what economists would now term a "country on the point of take-off", and it rapidly caught up the greater part of the deficit. This process was accompanied by a revolution in social structure: the traditional skilled labour force of the classical northern industrial centres was joined by a mass unskilled labour force - a generation of workers new to industry and recruited through internal migration.

The labour force and its political and trades union organisation had emerged from the war as a powerful, self-contained entity, a product of its stance of resistance to the fascist regime. Yet the organised labour movement was condemned to oblivion for a considerable period. This occurred for various reasons, including the exclusion of the Communist party from the centres of power as the Cold War set in, the politically motivated split of a unified trade union movement into rival unions aligned to different political groupings, the development of the economic miracle, and the fact that the unskilled labour force was constantly replenished by the proletarianisation of the agricultural population. There was in fact one unique opportunity for revival, and this was taken up. The unwillingness of capital and the inability of the state to promote social equality and to modernise the institutions of collective bargaining combined to produce a dramatic change in social relations in the "hot autumn" (autunno caldo) of 1969. The essentially new system of industrial relations which resulted can only be understood in the context of the rich theoretical, cultural and class-struggle tradition of the Italian labour movement. The trade unions embarked on important processes of reunification and established themselves as a bargaining force which could no longer be ignored on political, economic or plant level.

3. Regardless of these national variations, the relative states of retardation and successful

efforts to make up the deficits have produced essentially identical technological situations. In all cases socioeconomic structures have undergone a simultaneous process of adaptation. This can be seen in three significant trends: state intervention in the economy; the development of welfare state systems; and well-established unions which mediate between the labour force, capital and the state have managed to increase the share of national income which goes to labour. These have been the major social innovations of the post-war years, largely initiated by Great Britain (Berger, 1983). Since the mid-seventies, however, their validity has been fundamentally challenged by the onset of an international economic crisis and the end of the product cycle, which, based on the technology of the 1930s and the 1940s, was finally exhausted after a long phase of maturation (Mensch, 1979; Freeman, 1983). All advanced West European states thus saw themselves confronted with a dilemma. The search for a new developmental model was under way.

By far the most important development in this respect is the advent of microelectronic technology. For the first time since electricity was introduced into the industrial process, a new basic technology has arrived. This has such a broad range of applications that substantial changes in the technological basis of the economy and society as a whole are inevitable. A characteristic feature of the new technology is the fact that the attendant economic growth is not associated with increased employment levels. Indeed the colossal potential for rationalisation means mass worker redundancy. This has shifted the balance of social forces to the disadvantage of the unions and forced them onto the defensive everywhere. But more than the straightforward loss of economic and political powers of bargaining and assertion is involved: there is also a strong tendency towards erosion of industrial relations systems. Because the advent of microelectronics can be viewed as the start of a third industrial revolution, changes can be predicted in the political as well as social decision-making and steering systems. Traditional political and social actors could lose their hold if they fail to regenerate themselves. So far the main product has been the dissolution of outdated forms of organisation and representation, and no politically, economically and socially viable transformation model for a new stage in social development has yet emerged.

4. None the less alternative developmental models are beginning to take shape:

- The liberal-conservative scenario chills the spines of classical reformers and in particular the trade unions, because it denies them a future. This scenario is a product of the alliance (holy or unholy, depending on the political standpoint of the observer) of "social" conservatives and "economic" conservatives. Social conservatives are keen to re-establish a strong state which would keep the unions at a distance from the political market. Economic conservatives would like to steer the economy by means of the most unregulated market mechanisms possible and, above all, to eliminate the trade unions as obstacles to the free market. For this doctrine, authoritarian solutions enacted by the state and the market take precedence over co-operation and intervention.

A leading conservative theorist predicts that the new dynamics of high technology will trigger an international, world-market-orientated phase of industrialisation. This in turn will give rise to a superindustrial society which, by the beginning of the 21st century, will be able to control its problems. This is the point at which the real transition to the post-industrial economy will begin (Kahn, 1983). According to this viewpoint, the coming boom, which is expected to bring growth but no appreciable reduction in the current level of unemployment, can only come about if the old Keynesian model and its entire ideological basis are discarded in favour of a relatively non-interventionist system. In his opinion, the coalition of "economic" and "defence, national security and foreign policy" conservatives with "social" conservatives has the opportunity, for the first time in 50 years, of challenging the political and economic leadership of the old alliance of interventionists and "have-nots" with their constant refrain of social justice.

- But it is not only conservatives who view the future of the trade unions with scepticism (cf. Mueller-Jentsch in this volume). It is predicted that the dynamics of technical and organisational innovation in the production process will proceed without significant opportunities for intervention or control by the unions. Continuing high levels of unemployment are permanently undermining their bargaining position. Loss of jobs in the classical industries, where union density is traditionally highest, means reduced union membership. The unions, already threatened with losing their position as the recognised representative of the labour force, cannot offer a positive policy for the future, and will increasingly lose importance as a social movement. It is assumed that the identity of the unions is fundamentally wasting away, and that they have by and large already surrendered their capacity for renewal. In other words, it is suggested, time has run out for the unions. The most they could hope for would be to survive in a "dual economy" as representatives of workers who have relatively secure jobs. They would no longer carry any political or social weight in the development of society more generally.

The first scenario fails to take serious account of the inherent problems of the liberal-conservative strategy and grossly overstates the discernible developmental trends. In particular, it overlooks the fact that, while the technological, economic, political and social transformation process may well change the framework within which the unions have to function, it also opens up new room for bargaining. A model for the future which is based on the classical values of freedom, equality, justice and humanity is not merely a political dream: it can take strength from the knowledge that the conservative alliance will not be able to maintain the stability of its current hegemony, and when it eventually fails, alternative political models will be called for. A typical feature of social and economic conservative strategies is that they are based on the unequal distribution of political, economic and social burdens, to the disadvantage of the workers. The longer this process continues, the less convincing the liberal-conservative arguments will be. New processes of political education and mobilisation will be welcomed, preparing fertile ground for unions, classical reformers and new social movements.

In the context of the second, alternative scenario, the unions must first perform their traditional, protective role more consistently than in the recent past. The field of action here is broad. For the so-called "post-industrial economy" is still a long way off, and will presumably be preceded by a phase of total industrialisation of society. But the current union strategy of trying to resist the social consequences of technological change can be viewed as no more than a defensive response. Similarly, the goal of a more equitable social distribution of the wealth created by the new technology is not good enough. However important it may be to develop traditional union policy forms in line with modern demands, this does not meet the need to confront inevitable social change with alternative models of development and political forms (cf. contributions of Butera/Della Rocca, and Accornero). This is an intellectual task which the unions, classical reformers, and new social movements cannot perform alone. They must be involved in it, none the less, if the search for a future model of society is to be politically effective. Just as conservatives traditionally attempt to justify crises in terms of their purgative effect, so radicals are constantly searching for opportunities to renew and advance a humane society based on the solidarity of its members. Since the coalition of social and economic conservatives automatically generates opposition, a coalition of old and new social movements could well become the dominant political force.

5. Reviewing the past five years, it is easy to see - and the contributions in this volume provide a wealth of examples - that economic crisis and technological change have realigned the balance of social forces. The unions are at a disadvantage in the plant-level, economic, and political markets, while the leading liberal-conservative strategies, which are also executed in a diluted form by social democratic parties in government, make every effort to prolong the crisis and thereby promote its purgative effects. The division between the Keynesian model of state intervention and social consensus, and new developmental models is becoming increasingly apparent. Clearly the process of social transformation is only in its infancy. For industrial relations systems are almost exclusively displaying erosion tendencies, with barely any evidence of a new structural, functional or institutional order.

The following developmental trends are discernible in Great Britain, Italy and the Federal Republic of Germany:

- In Great Britain the trade unions have largely been excluded from the macropolitical arena because they lack economic market power and are subject to a government which opposes them. Both state and capital are now directing their efforts towards destroying the formerly strong power base of the unions at plant level. Whatever the importance of state economic policies and trade union legislation in this context, the economic crisis, unemployment, and world market pressures to adopt the new technology give management the chance to regain control of the production and labour process. The changes in production technology, labour organisation and industrial relations at British Leyland were admittedly exceptionally radical. None the less they provide a prime example of the capitalist strategy of engineering exemplary defeats of the unions in general, and the institution of shop stewards in particular, by means of authentic power struggles as opposed to simple reliance on legislation or spontaneous, "molecular" changes. The strategy of re-establishing the plant as the power base of the management is executed in various ways: the traditional organs of collective representation within the plant may be destroyed, excluded, by-passed or substituted by new forms of consultation. That there is currently considerable competition and conflict among and within the unions in Britain is more than a simple example of the divisiveness of union competition: it also reflects the way in which areas of conflict which were traditionally located outside the unions can be transformed into potentially self-destructive internal conflicts.

- Although less dramatic, developments in Italy are similar. The disputes at Fiat are a clear example of how modernisation of the labour process in its technical and social moments can also be used as a lever to restrict the broad powers implicit in the form of union representation within the plant which had been hammered out in previous disputes. Not only do the unions now have less power, they are also more willing to compromise, since they have a vested interest in maintaining a modern, internationally competitive enterprise in order to protect jobs. The Italian situation differs from that in Britain or West Germany. For the fact that the Italian unions are willing to compromise also characterises the very intensive tripartite discussions in the political arena. Arguments centre on the "scala mobile" (an index-linked mechanism to balance wage and price increases)is incompatible with the current economic system and the prevailing situation of international competition. The fewer concessions the unions gain in return for the compromises they make, and the more effort the state and capital put into reducing the effectiveness of the scala mobile, the more apparent becomes the mutual obstructiveness of dissent among and within the three union confederations. Erosion of industrial relations threatens to destroy the previously complementary effectiveness of decentralised and centralised union power from both ends.

- The erosion process in the Federal Republic of Germany is more insidious. Institutionalisation and legalisation of the social conflict of interests and the incorporation of the unions have been largely successful. Thus the state and capital have little reason to implement corrective or regulatory measures in a system of industrial relations which is already endowed with the capacity for self-regulation. The manner in which labour relations at Volkswagen have developed is unique in some respects but is by no means an exception. The almost imperceptible erosion of industrial relations might flare into open crisis if the state and capital are encouraged by the current weakness of the unions to renounce traditional co-operative relations with the unions and attempt to defeat them. With the alienation of labour force and union consciousness, and the first open split of the unions into two fractions, signs are growing that such a change in strategy is taking place. Thus management is now far less willing to offer concessions in order to preserve the social consensus. Likewise the conservative-liberal coalition has begun to intervene in current wage disputes to an unprecedented extent. The government is refusing to accept the demand for a shorter working week, on the grounds that it would be detrimental to the national economy, despite the fact that the demand comes from the fraction which includes the metal workers' union, the largest and strongest labour organisation in the country. And at the same time, the policies of the other fraction are being favoured under the terms of a new law which provides state funds for early retirement. Distinguished management representatives, however, are warning against an all-out attack on the unions. For they believe that it is in the interest of all to preserve trade union organisations which have external bargaining authority and internal authority over their members. Thus industrial relations in the Federal Republic of Germany could be said to be at a point of make or break.

6. In essence, the shift in the balance of social forces has run parallel to economic crisis and technological change. The state and capital have cemented their political and economic leadership and re-established their virtual monopoly of control over industry. This illustrates the degree to which

state and capital have successfully modified the positions of power which the unions had carved out for themselves in the post-war years. What is remarkable is the paucity of functional, structural and institutional innovation. While encouraging erosion in industrial relations systems, the prevailing liberal-conservative strategies fail to provide the basis for a new model of interaction. This is a characteristic feature of the first stage of the transformation process governed by the alliance of social and economic conservatives. It also demonstrates the problems of a model of political, economic, and social restructuring which attempts to return to the pre-Keynesian developmental stage of capitalist democracies. Discussion of the form of the future social development has only just begun, and the outcome is still open. It would be criminal to ignore the dominant tendencies emerging today; but it would be equally wrong to treat them as irreversible changes already inscribed in the new technologies.

REFERENCES

Berger, S., 1983: "Introduction", in S. Berger, ed., Organising Interests in Western Europe, Cambridge.
Freeman, C., ed., 1984: Long Waves in the World Economy, London.
Kahn, H. 1983: The Coming Boom, London.
Mensch, G., 1979: Stalemate in Technology, New York.

Part One

TECHNOLOGICAL CHANGE AND LABOUR RELATIONS

Chapter One

TECHNOLOGICAL INNOVATION, ORGANISATION OF WORK, AND UNIONS

Federico Butera and Giuseppe Della Rocca

This paper presents some hypotheses and data on technological innovation and its effects on the labour market and industrial relations. We believe that it is too early to present a definitive account of these issues for three main reasons: the nature of technological innovation has not been sufficiently explored; traditional categories of analysis do not deal adequately with the relationship between technological innovation and the labour market; and, thirdly, there is insufficient knowledge about the expectations and claims of new social strata and professional groups and/or their implications for industrial relations and union bargaining. Accordingly this paper draws on our earlier work to illustrate some of the trends at work (for details, see bibliography).

Industrial Automation: the Objective and Conceptual Revolution

There is much discussion in all industrial countries on the impact of microelectronics and information technology on employment, work organisation, and society. Indeed this is seen as one of the major economic and social issues of the 1980s.

As a result of low costs and widespread application in both productive and nonproductive areas, new technology has developed with an unprecedented speed. Some scholars maintain that new technology involves a rupture in technical development because of its "intelligent" function, science-based technology, and global application (Rada, 1980). But when the social effects of the wide variety of applications of microelectronics in information processing are examined, it is impossible to decide whether these result from the technology itself, its various spheres of application, or other related changes in production and society. Thus the idea of new technology should be incorporated into a more complex analysis which could account not only for the conversion techniques but also for the purpose inherent in the technology itself.

For the purposes of understanding changes in work and organisation, new technologies are best studied when they are components of a composite, higher level phenomenon called automation. Not all cases of new technology are employed in automation. Electrical, mechanical, hydropneumatic (etc.) technologies as well as information technology, are included in automation, and not all information technologies are used in automation.

We have evidence from various sources that automation is the continuation of an historical process in which human work is incorporated into machines, taking human work to be the mental and physical effort required to transform physical objects and information from one state to another, while at the same time exercising control over the technical and social conditions involved in achieving this transformation.

Automation is, however, a rupture in the historical process because the extent and speed at which work is incorporated into machines produces a qualitative change in the elements of the industrial unit, in the structure of industrial work, and even of society itself. Novelty does not lie in new control systems, computer technologies, integration of production flow and integration of business functions taken individually, but in all of these taken together.

We think automation is an appropriate word to describe a specific phenomenon. It is the displacement - within the physical system itself - of goals, models, procedures and languages for controlling and integrating meaningful conversion processes. Thus not only work but typical social processes/structures enter the machines: decision making, values, languages, power structures, co-operation and so on.

We are faced with a revolution. Almost any sort of work content can now be performed by automatic devices, any production process can be automated, firms of any size and in any sector can be involved in automation because of the availability of microprocessors and the sophistication of EDP

(Electronic Data Processing) applications. Automation never consists of "production systems without men", but involves production systems where men choose, implement, modify and run a highly formalised joint technical and social system.

Automation is a new phenomenon; thus the analytical categories for forecasting and evaluating the social effects of technical change must be reviewed. Certain basic social characteristics of the present industrial system could also be challenged, such as full employment, preservation of trade skills, grade differentiation, blue and white collar differentiation, apprenticeship training, career, long working days; some traditional elements of work design and negotiation procedures might also prove to be ineffective.

This scenario will change the rules of work design and industrial relations. Yet it is impossible to forecast the impact of new technology until new concepts and categories of industrial organisation and industrial relations have been established. This involves conceptualising a large variety of <u>project areas</u> or <u>game-tables</u> dispersed along a macro-micro continuum. Due to the rate of innovation, such "project areas/game-tables" display different features:

- Each of them has different objects, techniques, goals.
- They are not hierarchically ordered from macro to micro but have systemic relationships: innovation in one area may change the rules of the game for the others: a new design for the rules governing the labour market may alter, for instance, the policy of technological design; likewise, the development of a new micro-chip in a small laboratory may alter the international division of labour.
- Subjects strong in one area/table may be weak or absent in others: technicians as social subjects are strong only at the plant level, employers' confederations are strong in employment regulations but not in the design of new plants, etc.

Most of the paradigms and models offered for interpreting relationships among labour market, production techniques and industrial relations fail to recognise the new features of the present industrial revolution. These include: the increase of relevant structural elements of the socioeconomic system, their nature as <u>project areas</u>, their interdependent systemic relationships, the increased opportunities of choice for a growing number of subjects organised not only within corporatist groups but also within the project areas. An interdisciplinary and participatory way of designing complex systems is emerging: it also involves an increasing number of decision making/negotiation "game-tables" different from those found in centralised or corporatist models.

The empirical evidence concerning the present rules of the game in work design, institutional design, and industrial relations may be altered in the near future in so far as project areas develop. Thus the birth of small enterprises, the organisational changes in large firms, the new educational systems, the new plant design, the experiments in participation and so on may reverse - for better or for worse - any forecast we may presently make on issues like polarisation/recomposition, downgrading/upgrading of work, centralisation/decentralisation in the firms, collaborative/antagonistic industrial relations. Choices and decisions take command in the new technological revolution. We strongly disagree with those who see these choices and decisions concentrated in the hands of only a few subjects, or as depending on a few critical variables.

Automation and the Design of Jobs and Organisation

Evidence for the preceding thesis is offered by the recent ILO research on automation and work design in 13 Western and Eastern countries (Butera & Thurman, 1984). There are four basic types of industrial automation. Each type encompasses various degrees of sophistication of computerised control, production integration and uncertainty in the unit.

Type I, called "electronic mechanisation", includes isolated computer-assisted machines. It intensifies the problems typical of traditional mechanisation in work and organisation. Human tasks basically become ancillary to the machines. Different designs of technology and different ways to combine tasks are possible. Serious problems of quality of working life, worker polarisation and system disruptions are likely but can be avoided by various means.

Type II is "full automation" in a fully formalisable production process, similar to the ideal of an "automatic factory", such as a power plant, oil refinery, etc. Human tasks are removed from the actual transformation process and passive monitoring prevails. A contrast arises between the tremendous responsibilities for errors and the limited ability required. In order to avoid the dramatic consequences of lack of attention, alternative designs have been reported in EDP, instrumentation, roles, work groups, and education.

Type III - "incomplete automation" - includes the majority of automation cases. Both machines and the production process are run automatically but disturbances are frequent in input, conversion, or output, as in paper mills, printing factories, aircraft automatic pilots, etc. Uncertainty requires that the information loop be closed through people and that task interdependence be improved.

Type IV is similar to III but there is also a large amount of manual work to do, as in an automated rolling steel mill. People must have full control tasks on the process and - unlike Type III - cannot be remote from the process itself.

All four types permit organisational choices which give workers skills and control. Types I and II may clash with conventional bureaucratic/Tayloristic organisations, mainly because they may induce poor motivation, unreliability and errors. The organisational requirements of Types III and IV are more inconsistent with traditional work organisation. In order to function they require an organisation where the automated control system is under the continuous control of a human control system.

Tasks are the building blocks of work organisation, the interfaces between technology and organisation. Tasks therefore impose certain functional requirements on the work organisation. Thus, if the organisation does not meet the requirements of technology or does not take advantage of the freedom it offers, dequalification, poor quality of working life, negative attitudes and system failures may arise. The breakdown of the Three Mile Island nuclear power plant is a well-known illustration of this (see Fadem in Butera and Thurman, 1984). Control tasks in automation have some general properties that have a direct effect upon work organisation. These requirements are: a) to absorb a limited amount of workers' time; b) to respond to a "stochastic" event of importance, i.e. to be accountable for possible delays in taking immediate action at an unpredictable time; c) to involve a high level of functional interdependence and communication among the different actors; d) to demand patterns of activity which vary on each occasion according to productive and organisational events; e) to require an extension of knowledge beyond the most frequent activities performed, to the point of understanding some of the "laws of the process".

These properties require work to be designed differently. There must be a move away from the traditional design which includes strict job specifications, full employment of work time and measurable activity, clearly defined boundaries between jobs, the concentration of information and co-ordination in supervisory roles, the segmentation of different activities; in the traditional design there are different job titles held by individuals with different training and wages, and limited vocational training.

On the other hand, the design of such elements as jobs and work groups, organisation units, occupations, rewards and so on, is typically a "social invention"; they are also influenced by immediate social factors such as legislation, industrial relations, ideologies, values, etc. Thus automation has no deterministic effect upon the final structure of the organisation. The same can be said about the quality of working life and occupations. The chains of influence stemming from the requirements of the productive system (automation -- tasks -- organisation -- quality of working life) meet up with those stemming from the social system (institutions -- social action and behaviour -- organisation -- quality of working life).

Two main types of "manual" workers have been discerned in empirical research into automated settings: the "ancillary worker", who is in charge of supplementing the imperfection of automated equipment, and the "process control worker", who has actual control over technical processes and relative knowledge and status. Naturally the first type can be found more frequently in automation Types I and II, while the second is more often associated with automation Types III and IV. Purposeful design can create process control workers in any situation, while incorrect design can create ancillary workers where this is totally inappropriate.

Most of the reported cases of polarisation of work were found in Type I automation with a few workers overskilled while the majority is given simple, repetitive, unskilled jobs.

Automation of Types II, III and IV imposes functional requirements on the recomposition of work: when jobs, organisation units, communication and decision-making processes are fragmented they are no longer consistent with production objectives. Recomposition of work and knowledge, work and organisation, work and life is also required.

If technological functionalism could be explanatory and predictive, the functional requirements of Types II, III, and IV of automation should lead to a recomposition of the occupational structure, i.e., the elimination of unnecessary differentiation among workers regarding job titles, grades, wages, etc. Evidence of this impact is, in fact, found in the smaller number of job titles, reduction of grades, full task rotation, new extended production professions (for example the process operator), and the blurring of sharp distinctions between blue- and white-collar workers. But in the same technology we can also see the opposite phenomenon, the so-called "polarisation" of the work force.

The different types of automation may partly explain this phenomenon: Type I being more conducive to polarisation and Types II, III and IV to recomposition. But there are more important societal factors to consider. The trend towards segregation and segmentation of the work force is still taking place within the labour market and is reinforced by increased unemployment. Old social patterns and technological unemployment lead to a polarisation of the occupational structure which contrasts with the functional requirement of work recomposition that may arise from most cases of automation.

This has the following implications:

- Planned design of technology and organisation with joint and anticipated consideration of the social and economic dimension is crucial. Interdisciplinarity, participation and consideration of the unique character of any organisation are components of good methodologies.

- It is possible to indicate some potential and plausible principles for producing a new paradigm of work in automation -- principles which have been seen in operation and are judged to be generalisable:

i) jobs should include tasks of EDP control, direct control and manual tasks, avoiding idle time and encompassing procedures requiring continuous action;

ii) role design should be based upon functions performed, results achieved, variances absorbed, interactions: most roles will include not only conversion but also co-ordination, maintenance and innovation;

iii) tasks should be rotated;

iv) work groups should be self-maintaining units where people can express their roles and co-operate;

v) new professional patterns should be established, different from both fragmented jobs and traditional trades; they will include skills to supervise computerised control systems and actual physical processes; they will not be linked to a specific plant or firm (process operator);

vi) training in theory and practice (algorithmic versus heuristic knowledge);

vii) short work hours;

viii) continuous career, reducing the gaps between the blue collar workers and technicians.

- Other paradigms may emerge from different compositions of other economic and social variables and from choices which ultimately depend upon knowledge, power and emotions.

Technological and Organisational Change in Italy

The acceleration of technological development in both large and small enterprises is a driving force of organisational change.

i) The entire range of substitution of transformation technologies has been experienced in Italy, from more efficient small and large mechanical machine tools to new chemical processes, etc. Numerical control machines, machining centres, flexible computer-based manufacturing systems are widely adopted in small units as well. Robots in assembly, welding and painting appear at an increased rate and computerised control is the norm in process industries. Most of these technological innovations, unlike in other countries such as the F.R.G. or Japan, are introduced on a scattered, day-by-day basis, according to a "small steps" policy. Most of them are labour saving and reduce the transformation skills required of the blue-collar workers. Because of the prevailing adaptive pattern the technological innovation is often developed in co-operation with the builder and the user.

ii) The second area of technological innovation concerns transportation technologies, material handling, buffer systems, etc. New ingenious solutions are applied and a general computer-assisted system approach to the physical displacement of material is often developed in various industries. Savings in the work in progress and higher manufacturing flexibility are pursued. A strong impact is often noted in the relationships among human tasks, so that the structure of work relationships and, consequently, the content of roles are modified.

iii) In Italy, as elsewhere, Data Processing technologies show the most striking changes. The country's ratio of introduction of Central Process Units (CPU), small special-purpose computers, peripherals, network and software packages, etc. has been very high in all the areas of industrial application such as management administration, sales, manufacturing, R & D; and office automation is now appearing. The boundaries of EDP philosophies and applications often cross those of organisational philosophies and projects.

The main research issues are centralisation versus decentralisation, concentration versus diffusion of control on information systems, open versus closed loop feedback on processes, design of jobs encompassing understanding of the control cycles of events versus design of jobs ancillary to the EDP, etc. Issues like participation in EDP design and joint design of EDP and organisation are widely discussed (see next paragraph).

iv) Finally, innovations often involve "systems technologies", i.e., complex and homogeneous systems of machines, information systems, methodologies, control systems, integrated business function through a total system design. Many managerial approaches to decision making, such as Management Decision Systems, are largely computer-based.

In summary, we can say that Italy does not have specific programs, government activities, established institutions dealing directly with the quality of working life, humanisation of work, humanisation of technologies and so forth. There is a profound process of modification of the industrial systems, where management, unions and government cope - in various ways and contexts - with serious economic and social problems, where numerous experiments and learning activities on work organisation are underway, where an emerging informal network of professional and non-professional students of work re-design is very active and innovative (see next section).

Organisational changes in Italy are difficult to survey quantitatively. Some main characters may be pointed out.

1. Organisational change is intense in all industries, whether affected by technical

development or not.

2. In some cases, organisational changes are associated with growth of turnover and employment; in most cases (e.g., steel, shipbuilding, textiles, chemical fibres, etc.) with the decline of both. In other cases (e.g., the auto industry) only with a reduction in employment.

3. The economic crisis put pressure on productivity making some management practices acceptable that were rejected ten years ago: strict control on absenteeism, incentives, bonuses to middle management, tightening of manning, etc. In some cases (mostly white-collar employees and technicians) we have Tayloristic fragmentation. Managers for the most part do not believe in the traditional rationalisation based upon the intensification of direct work.

4. The introduction of new technologies is associated with opposite types of work organisation: CNC with high polarisation of the work force and the recomposition of maintenance and programming; CAD/CAM where designers are split into two groups and where new creative duties are given to workers; cases of process control operators prevented from understanding the processes, and cases of re-designing the control room and softwares in order to keep control within the workers' capacity, etc.

In most cases a new organisational paradigm emerges in industry which moves away from the models of the '50s and '60s (Butera, 1984). In the big industries the process of "making large small" is visible: for example, decentralisation, divisionalisation, etc. Self-contained units are goal-orientated. Control systems are more sophisticated where both central and local control increase through distributed informatics. Middle management is used in controlling results, not people. Temporary and matrix organisations abound in dynamic enterprises, breaking the "dogma" of the unity of command. Partially self-regulated work groups are diffuse. Jobs may be "rich" or "narrow", but work roles are frequently identified by results, relationships, actual control of variances.

Technological/Organisational Innovation and the Union in Italy. The Change in Industrial Relations at the Enterprise Level

The problem of participation in the design of the applications of technological innovations has become particularly significant in the case of information technology.

Unlike other technologies, the debate has been provoked not only by external pressure such as the action of the union organisations or workers organised to protect themselves from the consequences of the innovation and to safeguard the quality of working life but , mostly - at least in the initial phase - by internal pressure, due to the needs of the design process itself. It is no accident that a fair share of the debate developed in technical areas and was carried on by computer and organisation specialists.

The specific nature of information technology explains the significance of the design process. It has a dual level: the choice of hardware (and of the related basic software) and the development or purchase of the applications software, i.e. the programmes that serve to automate a specific procedure. In the design phase, before the development of the programme, information technology is less binding; once completed, it is generally more rigid: the greater the increase in automation, the more the methods of organisational functioning are bound to the way in which the automation itself was designed. In other words, the flexibility is fully contained in the design and it must be forecasted completely in it (De Maio et al., 1982).

At the same time, it is the programme that defines the margins of autonomy or self-determination within which a person has the faculty to choose the objectives to be reached with his work, the alternative means for reaching them, and the external variables to be taken into account.

Thus, the results of an application of information technology - in terms of both technical-economic performance and the consequence on the quality of working life - are greatly conditioned by the design process. By means of participation (of the users) in the design process, the system may be made to perform better. And it is only through participation in the process that it is possible to condition the results on the quality of working life.

As Crozier maintained, with the entrepreneurs' demands to participate, we see a reversal of perspectives. "For reasons of efficiency, the management asks the employees to participate and, to obtain the participation, it tries to transform the rules of the game that govern the employees' working life. This implies that participation is no longer considered a gift from the management to the employees, but as a burden to impose on them, for which they should be compensated" (Crozier, 1973). Through the participation of the employees and their demands the union finds itself obliged to enter these processes or at least to make a decision in this regard.

Unlike other countries, especially those of northern Europe, in Italy the issue of the technological innovation of production remains unresolved in the industrial relations area. Not only do the parties contest rights and prerogatives - a common occurrence throughout the industrialised world - but there is also questioning of any form of union participation. Thus three alternatives are being considered in Italy. The first of these is to abandon bargaining in the area concerning the organisation of work. Here, the union must limit itself to bargaining issues outside the realm of production, mainly income distribution. This was the thesis of the Federmeccanica (Italian Metal Trade Association) in 1981 and it is found in some of the examples mentioned above. But it is also the thesis of some union sectors and of those Italian intellectuals who advocate political exchange and neocorporative-type industrial relations.

The second alternative is the reconstruction of union rights in order to protect the workers from the negative effects of technical-organisational innovation. It is seen in the unions' proposal for defensive control: i.e. reconstructing a policy of claims that can restore the "lost rigidity", bargain salaries, working hours and so forth in the face of every innovation. In this case, the union is given only the right to negotiate the effects, the end result of the innovation. This alternative does not imply a change of the actual industrial relations system.

The third alternative seeks anticipatory bargaining and participation orientated towards technical-organisational innovation. The union participates and thereby not only establishes rights of protection but also helps to fix the objectives and orientations of innovation and the rules for its attainment. This alternative presupposes a change in industrial relations and requires not only that the parties conduct negotiating-conflictual activities but also those of a consultative-participative nature. In other words, there must be a change in the pattern of industrial relations, predictability of conflict, and consultation and experimentation rather than bargaining and conflict (Della Rocca & Negrelli, 1983).

Some of the research on bargaining at the enterprise level shows the prevalence in 1982 of the second approach in spite of the affirmation of the rights to disclosure of information in the 1977 category agreements.

In actual fact, the rights to disclosure of information involve several principles, such as that of formally introducing into bargaining activity the problems, not just the information, of the enterprise. Secondly, they partly modify the bargaining procedure, establishing, either formally or informally, a consultative phase. Nevertheless, the rights to disclosure of information are not always used as anticipatory information for an enterprise decision and for negotiations. And there are very few agreements which establish a joint operational-managerial structure for the processes of technological innovation. So rights to disclosure of information do not always mean real participation and mutuality. Three categories of usage of these rights have been identified:

- constrictive, in this case the function of the rights to disclosure of information is often one of closing off any type of negotiations. It is no different in practice from the simple communication to the union organisation of decisions already made by the enterprise;
- compatibility-based, the most frequent category for disclosure of information. It concerns those cases in which these rights introduce both a general (market, investments, etc.) and specific (labour mobility) framework of compatibility. Here the information serves both to establish the bargaining constraints and to constitute the vehicle through which the entrepreneurs present their demands to the union. This category allows an exchange, and the bargaining considers the constraints of the enterprise first and foremost;
- participative, regarding those agreements on the rights to disclosure of information which promote a form of participation for the verification and discussion of the general or specific objectives of the firm (investments, market, employment, training, technological innovation) and which give rise to a set of constraints that are not only economic/productive but political/institutional. Consequently, these agreements are strategic in nature and establish a preventive-consultative type

procedure with instruments of joint management. Only here may the use of the rights to disclosure of information be considered as a change in the previous industrial relations models.

Which type of application prevails depends on the evolution of bargaining and on the attitudes of the parties in particular. The communication of information on the state of the enterprise is not a new practice; firms, especially during periods of crisis, have always used the disclosure of information for negotiation purposes. The formal recognition of this practice in the national collective labour contract has codified the union's rights to disclosure permanently. Yet this right remains basically procedural and its results depend on the parties' behaviour and the use they make of it. This explains the different types and results identified in the analysis of enterprise bargaining over the last five years in spite of the fact that these rights were interpreted by union organisations as an initial means of participation and control on what are and were considered prerogatives of the enterprise.

Studies of enterprise bargaining show a change in the contents of bargaining: a change that occurred during the crisis, after 1975, which meant that bargaining activity was more dependent on the firm's economic/productive management. Thus the rights to disclosure of information were basically used within the area of the firm's market compatibility, and their effectiveness was limited by the bargaining results connected with the productive performance.

The research also shows that union intervention concerning the technologies and organisation of work may occur where there is a relational-participative type orientation. In some cases this type of bargaining leads to agreements on certain medium-long term aims. This means the agreements are strategic and planned. Here the aims are not defined exclusively in terms of economic and market constraints but also in terms of social and political-institutional expectations.

Thus the bargaining results of the past few years reveal a shift in negotiating principles. They are not openly stated by the parties but are implicit in their orientation towards the negotiations. The process by which this occurs is inductive more than deductive; the agreements are not general political-institutional ones but are initially experimental agreements on particular aspects which, after a certain period of time and once achieved, are gradually generalised.

The rights to information disclosure have also introduced changes into bargaining. The first and most important innovation is that these rights enable enterprise and union to forecast their behaviour in both the short and medium term. The change therefore comes from the shift from a basically unpredictable situation to one in which the behaviour of the parties is known. The diffusion and knowledge of the information plays an essential strategic role here.

The second innovation introduced by these rights is the anticipatory discussion of many of the changes provoked by the entrepreneurial strategies and management. This concerns in particular the technological changes and organisational restructurings. These agreements are not as widespread but they do guarantee the union greater control and capacity of realisation.

The third novelty comprises the development of informal union-enterprise relations. The broadening of the procedures by means of preventive information establishes a period of consulting activity which precedes the actual negotiations. In a conflictual situation where every aspect is rigidly underwritten and formalised, rights to disclosure of information may introduce a consultative phase that does not necessarily end with some sort of bargaining activity and yet still precedes it. Consultation does not formally involve the parties; they are not obliged to take explicit stands as in negotiating. It may introduce a phase of reflection and specific focus on issues of enterprise management. This informal activity does not impede a more complete and mature development of real bargaining activity.

The fourth novelty concerns the changes that the rights to disclosure of information introduce in the modus operandi and structure of union representation within the firm. The Workers' Delegates Committee remains a strong organ of worker protection and at the same time of consensus and participation in the organisation of the union (think of the level of anomaly and malfunctioning in industrial relations in those sectors and firms where the Workers' Delegates Committee is absent). Nevertheless, the changes described above lead to specialisation and division within the activity of representation, first and foremost among the different levels and subjects of the bargaining. With the change in bargaining contents three types of representation are taking shape:

- that of conducting a bargaining activity in the real sense of the term through negotiating, with recourse to conflict and whose result is a set of rules protecting the workers at the productive and/or firm level. These activities are conducted in the first place by the co-ordinating organs of the union representatives.
- that of conducting an activity of control and verification on the application of the bargaining norms, i.e. both those produced outside the firm and undersigned by the unions and those effected by the shop stewards. These activities are carried out by the shop stewards.
- that of conducting an activity of consultation and discussion without reaching a contractual agreement on the issues both directly connected to working conditions and those regarding managerial and entrepreneurial activity. This type of activity is conducted in the first place by the senior shop stewards.

Cases of Participation in Design of Work Organisation and of New Technology at the Shopfloor Level

Even though participation in design of work organisation at the shopfloor level has not been institutionalised, the past ten years have been rich in experiments of worker and union participation in design at the plant level, even prior to the agreement on "information disclosure". No common or prevailing procedure has yet emerged.

a) Most cases from the '70s involve the redesign of the organisation: unions and workers have been actively involved for the most part in aspects like job design, formal qualifications, employment conditions, working conditions, training (Olivetti, Fiat, Dalmine, Italsider and many others). There are a few cases of new design, where unions have participated in designing a new unit or in introducing new technology. The best-known examples are in the steel industries: Italsider and the new Dalmine plant. At present, when new technologies are introduced (EDP Application, Robots, CNC, Office Automation, CAD/CAM, etc.) there are cases of consultation with employees but not with unions, unless problems concerning union prerogatives emerge (employment, grades, training programmes, etc.) or they are associated with company crises (e.g.

Italtel).

b) As far as the initiators are concerned, in the first half of the '70s most projects were initiated on the basis of a more or less formalised demand from the unions: from general claims in the Fiat case, to the binding claim for 2,500 upgradings at Olivetti, to the inclusion of re-design projects or action-research in formal agreements (Mondadori, Honeywell, Peg, Anic, etc.). Whenever management reacted to union demands nothing happened. Changes occurred when actual problems of the enterprise were acknowledged. The union had more autonomy in "Medicina dei lavoratori" for health protection but the subject of design/re-design was often limited.

In almost all cases of unlabelled change and in most cases of projects in the latter half of the '70s, industrial management was the initiator, calling in workers or management. In most instances the suspicions of unions and Workers' Delegates Committees were very great when the proposal for participation was on management's ground (productivity, organisation structures, organisational research, etc.). Famous examples here are the early projects in the public steel industries, which were substantially boycotted by the unions. When the projects were incorporated into a bargaining scheme (Italsider, Pirelli, Alfa Romeo, Ansaldo, Aeritalia) there seemed to be much more understanding among the parties.

c) The typology of participants in these projects is quite substantial. When the dominant participants have been the top management of firms and unions (Olivetti, Ansaldo, Pirelli, Alfa Romeo), the projects have more frequently led to formal agreements. A fundamental role has been played in some projects by middle management and delegates. Those projects were more successful as learning processes (Mondadori, Terni, Honeywell) but were not always productive in terms of concrete results and industrial relations agreements. Direct participation of workers has frequently been neglected or adopted in a manipulative way (some socio-psychological approaches ran that risk), or resulted in "hidden prebargaining", or in management's attempt to speak to work groups instead of the unions doing it. On the other hand, unions have been very slow in understanding the potential of direct participation.

Direct participation was seen as a potential disruption of present industrial relations. Now "Quality Circles" will be diffuse because they reach the same goals of direct participation while remaining totally detached from industrial relations. Particular examples of widespread worker participation are those of "training-research", where the analysis concerned both educational software and input for change (Keller, Italtel).

d) participation is practised in different ways in the various stages of a project´s development.

Diagnosis is the phase that union and management are more accustomed to sharing: what the situation is, what to do, etc. It can take the form of antagonistic or collaborative relations. The rights to disclosure of information are mostly a procedure of joint diagnosis.

Goal definition means to define, bargain and agree upon the concrete goals and parameters of the project. Two main difficulties have been encountered in this phase of the mutually independent definition of the "game rules": vague and ideological definitions of goals together with difficulty in operationalising the twin goals of organisational effectiveness and Quality of Working Life. This phase could potentially.
conventional bargaining and institutional procedures, but not with traditional ideological wording in industrial relations.

Analyses have taken different forms and have been undertaken by management or consultants and diffused after the fact; shared with Works' Councils; conducted jointly by technicians, middle management, delegates, workers and external researchers. Extensive participation of workers and unions in the analysis has been a peculiar trait of analysis procedures in Italy and therein lies the innovation in sociological and organisational studies in Italy.

There are many instances of participation in the design of the architecture of the new organisation: models and principles have often been proposed by workers and unions, unlike in other countries. The border between architecture and ideology/story-book was often narrow. However, this has been a phase of possible union/management co-operation in design without undermining their prerogatives.

Detailed design has been the area of technicians and middle management. Proposals for union participation in this stage has been rejected at some point by both management (as a violation of management prerogatives) and unions (as potentially backing management decisions).

Implementation and experimentation. Participation in this stage was easy for workers as well. Risks of manipulative involvement were present when the design of experimentation (evaluation, training, follow-up) had not been carefully executed.

In summary, management and unions in Italy have gone through a large variety of participation exercises but they have often failed to recognise the different relationships (antagonistic, bargaining, institutional, collaborative, etc.) permitted or required by the different phases of design. In addition, the technical specificity and complexity of the various phases have often been misunderstood by both.

e) A large variety of technical approaches have been adopted within the participative approach. Mixed groups of workers/middle managers for analysis and design (such as the ergonomic experiences in Italsider), the "esperto grezzo" approach (Medicina dei lavoratori), action-research (Terni, Dalmine, Mondadori, Anic), research-training (Innocenti, Keller, Italtel), experimental follow-up (Olivetti), experimentation (Sogeda), etc.

In conclusion we may say that the above-mentioned cases demonstrate that each phase requires from the unions skills and procedures which are usually different from those required in bargaining and institutional action.

Another basic problem of participation proves to be how to satisfy different requirements: institutional participation requires universal principles and societal procedures rooted in political processes; conventional bargaining is based upon the zero-sum "rules of games" of industrial relations; organisational participation is contingent, rooted in the uniqueness of the organisation, and is based upon "design", that is a socially interactive invention of desirable and possible futures.

REFERENCES

Bartezzaghi, E. and G. Della Rocca, 1983: Gruppi professionali e sindacato nella progettazione delle nuove tecnologie informatiche, Fondazione A. Olivetti.

Butera, F. 1982: Impact of new technologies on work organisation in production, in: L. Beckemans, European employment and technological change. European Centre for Work and Society,

Maastricht; Italian translation "L'automazione industriale e il futuro del lavoro operaio" in: Studi Organizzativi, n. 2.
Butera, F. 1983: Worker participation in design and redesign of industrial organisation: notes on the Italian case, Friedrich-Ebert-Stiftung, Bonn.
Butera, F. 1984: L'orologio e l'organismo, Milano.
Butera, F. and Thurman, 1984: Automation and Work Design, Amsterdam.
Crozier, M. 1973: L'influenza dell'informatica sul governo dell'impresa, in: Rositi, F., a cura di, Razionalità e tecnologie dell'informazione, Milano.
De Maio et al., 1982: Informatica e processi decisionali, Milano.
Della Rocca, G. and S. Negrelli, 1983: Diritti di informazione ed evoluzione della contrattazione aziendale (1969 - 81), in: Giornale di diritto del lavoro e di Relazioni Industriali, n. 19.
Rada, J.F. 1980: The Impact of Microelectronics. ILO-Publication, Geneva.

Chapter Two

CHANGING SKILL REQUIREMENTS AND TRADE UNION BARGAINING

Zissis Papadimitriou

Changing Working Conditions and Bargaining Policy

Until the recession in 1966/67 the chief target of collective bargaining policy was to implement higher wage demands and a reduction of working time. Items like technical progress and its effect on working conditions and work requirement were of course topics of discussion on the agenda of trade union conferences - especially in the IG Metall (metal workers' union) - and were considered in the resolutions of annual trade union conferences. But none of this had practical consequences for the development of trade union collective bargaining policy and organisation policy. This one-sided collective bargaining policy, which was primarily concerned with protecting existing living standards, social status ("Besitzstand") and a reduction of working time, led to neglect of two important aspects of work, namely, effort and skill. When the intensification of efficiency standards and the deterioration of the work situation in the companies were discussed, this occurred mainly from the angle of wage policy. The trade unions' attitude in principle tended to favour money compensation for hazards and stress. This wage policy and efficiency policy could only be realised as long as the economy was prospering and companies were ready to make material concessions by granting higher wages and special bonus schemes. The employers' preparedness to compromise diminished, however, with deteriorating conditions for realising profits and with decreasing growth rates. The concomitant increase of pressure on the standards of performance in the companies together with a deterioration of working conditions (loss of skills, loss of meaning of activities on the shop floor, monotony of work etc.) have provoked defensive actions by the workforce. Strikes, fluctuation, absenteeism, restrictions of effort, etc., were the most common forms of collective and individual resistance on the side of the employees. Employers have reacted to this "crisis of motivation" - as it was analysed by research scientists - by developing defensive strategies which were primarily concerned with re-integrating the employees into the work process. As a result of this "motivational crisis" among the employees, companies have started to develop strategies of work organisation which could again secure their preparedness to increase performance and thus productivity.

On the side of the trade unions these developments - together with the trend towards economic recession which has rendered high wage demands as compensation for deteriorating working conditions impossible - have led to the inclusion of effort bargaining into their collective bargaining policy. The demand for elimination of restrictive working conditions has recurred at several trade union meetings and conferences.

The trade unions only turned away "from the idea of compensation for a lack of job-satisfaction" (Ohm/Wenk, 1981, p. 70) after the strikes in the metal working industry of Nordwuerttemberg/Nordbaden in 1973. During the conflict about the collective agreement on working conditions in the metal working industry, the pressure resulting from the strikes was instrumental in extensive demands for the organisation of working conditions and work requirements which could thus be secured by collective agreement (Brumlop/Rosenbaum 1979, p. 264 and Kern 1979, p. 121).

The adoption of the skeleton wage agreement II ("Lohnrahmentarifvertrag II") opened a new phase of trade union bargaining policy. For the first time, the unions succeeded in agreeing a rest-period of five minutes per working-hour and the trade union achieved improved protection against dismissals, a protection of wages, more openness in the management's establishing of norms and standards of performance, a ban on extremely short cycles in assembly-line work, etc. Regarding the organisation of work on plant level, the skeleton wage agreement II includes a regulation saying that "work place, work organisation, and work environment have to be arranged in a humane way." This regulation concurs with the articles 90 and 91 of the Works

Constitution Act ("Betriebsverfassungsgesetz"). A certain degree of co-determination is conceded to the works council with regard to the organisation of "work place, work process and work environment") which is, however, worded so vaguely that it admits different interpretations (see Vetter, 1974).

Articles 90 and 91 of the Works Constitution Act and the regulation concerning the organisation of work in the skeleton wage agreement II can be seen as the forerunners of the "Quality of Life" reform programme of the SPD/FDP coalition government. The central point of this programme is the claim for a humane organisation of work. "The 'debate on humanisation' was here (in the Federal Republic of Germany, Z.P.) above all, part of the reform programme, propagated by the Social Democrats, which promised - under its heading 'Quality of Life' - an extensive improvement of industrial working conditions. The trade unions resisted capitalist rationalisation measures with their concomitant impact on the working conditions of the employees, by attempting to gain an increasing influence on the standards of performance and the direct organisation of working conditions and by using political and collective bargaining regulations as an instrument of their policy" (Brumlop/Rosenbaum, 1979, p. 265).

Whereas the significance of this "humanisation programme" for collective bargaining had always been controversial, there was a clear change in direction after the change of power in Bonn in March 1983. Since then it has developed entirely into an instrument of entrepreneurial effort policy.

The different regulations as well as the instruments for organising work in a humane way in this context have been outstripped by later developments of technology. For the trade union conception of eliminating restrictive working conditions in industrial production is mainly based on experience drawn from mechanised production processes. Thus it cannot be applied to high technology computer-based production. The development of more adequate strategies of work organisation and skill protection, which take account of future developments as well as the interests of the employees, calls for a careful investigation of the development of skill under conditions of computerised production.

Technical and Organisational Tendencies of Production Processes

Whereas well into the '70s computerised technologies came into use mainly in large firms, this situation has changed considerably by now. In a period of recession and economic crisis, more and more medium-sized firms but also small firms have switched over to these technologies in order to improve their competitive position. Technological developments, such as the miniaturisation of electronic components ("shrinking" computer equipment which needs less room), an increase in efficiency by further condensing operation, and, last but not least, the subsequent "shrinking" in price of the hardware and software have been instrumental in the massive penetration of the market by computerised technologies. Having already enhanced the employers' readiness for innovation, this process of new developments in the field of micro-electronics has led, with increasing competition on world markets, to a new wave of technical and organisational changes affecting the entire West German economy. The reorganisation of the process of industrial production and work has primarily been due to the contradictory requirements of production and market. Whereas cost considerations call for a concentration of the production and work process, fluctuating market requirements call for flexible production structures. On the one hand the rationalisation strategies of the companies have to take into account the penetration of the whole production process by the economy of time (co-ordination and integration of working time, material time and machine time on all levels of the work process, with the aim of an optimal realisation of all factors of production); on the other hand they have to consider the flexible organisation of production, with regard to market requirements. The reorganisation of production in the different sectors of the industry, according to the criteria of the economy of time, takes place "neither simultaneously nor with the same dynamic force and intensity. The targets, course and degree of reorganisation processes seen from the angle of the economy of time, are in the end determined by the economic, technical and material conditions of production, which are varying from industry to industry" (Benz-Overhage et al. 1982, p. 89). Thus the effects of this reorganisation of production and work process on the skill requirements are different in each branch of the industry.

The reorganisation of production takes place mainly with the aid of computerised technologies. Because of their special features (like bulk memory, programmability and problem-orientated processing and analysing of a larger amount of data and information within a very short period of time and last but not least because of their immense potential for flexibility), computerised technologies are an extremely effective instrument for the companies' rationalisation strategies. Computerised technologies differ greatly from former techniques of production and organisation and must be considered as a specific new form of technology. As to the application of computerised technologies in reorganising the processes of production and work, there are three important functions which can be distinguished: they can be applied as <u>information or organisation technologies</u>, as <u>control technologies</u>, and - contrasting with these two - as <u>product technology</u> (see Brandt et al., 1978).

As information and organisation technologies, computerised technologies - in the form of electronic data processing equipment and process computers - take over functions which were traditionally carried out by work organisation. The application of computerised technologies as organisation technologies - unlike conventional production techniques - does not serve to expand capacity. The function of computerised technology in the shape of planning-, operation- and process-control computers is to integrate the timing and to guarantee the continuity of the production process. Their special significance, however, lies in co-ordinating the requirements of the market and the economy of production. Thus they take over the function of anticipating requirements of the market economy with regard to costs and co-ordinate them with the requirements of the cost structure and time structure of the organisation of production. This function of co-ordinating diverging requirements, defines the specific importance of computerised technologies in their effect on organisation technology (see Benz-Overhage et al., 1982).

As control technologies, in the form of machine-control and process computers, computerised technologies take over functions of the conventional production techniques. Incorporated NC- and CNC-control systems are therefore used to optimise the employed machines and machine systems. They can be found as NC- and CNC-machines in the mechanical production as process computers and as electronic data processing equipment for dealing with orders in production control, in design and in production planning.

The application of computerised technologies in the products (electronic components, micro processors etc.) marks above all the transitional stage between electro-mechanical and electronic production in electrical engineering. As product technology, computerised technologies contribute - like the standardisation of products and parts - to a rationalisation of production techniques. The miniaturisation of electronic components and the universality of their application, together with the by now immensely increased condensation of functions, result in simplifying not only the production but also the assembly processes.

All things considered, the application of computerised technologies in the form of organisation, control and product technology means a new leap in quality compared with conventional strategies of mechanisation, semi-automation and methods of work organisation. On the one hand, this new quality becomes apparent when we consider the potential of computerised technologies which, under the influence of economic crisis, are co-ordinating strongly diverging requirements of market- and product-economy in a flexible way; on the other hand, it can be seen in the reorganisation and penetration of the production and work process by the economy of time. Whatsoever their functional application, computerised technologies are crucially shaping the material structure of the work process and are extensively changing the conditions of the application and the quality of human labour in industrial production. Above all, these changes can be discerned in the displacement of human labour (unemployment) and in the fact that the organisation of the capitalist process of production gradually dissolves the special features of human labour. Both tendencies can be related to the changed relationship between machine work and human labour (vergegenstaendlichte und lebendige Arbeit).

Impacts on Skill Requirements

We must now look into the effects of these technologies on the development of skills in industrial production. This will be illustrated by the application of computerised technologies in three industries of the West German metal working industry: the car-industry, the machine tool industry and the electrical industry.

In the car-industry we can observe the application of industrial robots in functions like automatic handling (as in automatic spot-welding) in electronic control systems in production (car body making). In mechanical engineering we see the substitution of conventional machine tools by NC- and CNC-machines, EDP-aided planning, production control and the introduction of CAD-systems in design. Finally, in electrical engineering, the focus is the change-over from fabrication and assembling processes of products with electro-mechanical parts to products with electronic parts. This investigation focuses on the connexion between innovations in product technology and the technical and organisational changes in the production and assembling processes.

Computerisation in these three sectors is linked to a far-reaching change in the structure of skills on plant level. Indeed there are four different types of industrial work which are altered by the application of computerised technologies or which emerge for the first time. These job-types are: unskilled repetitive work, which is engaged on only a small part of the production of the end product; semi-skilled automation work on the new machines; skilled automation work; and, finally, complex, abstract, mental work.

In the process of automation - which accompanies the application of computerised technologies - unskilled repetitive work, engaged on only a small part of the production of the end product, undergoes two changes. It is replaced by machinery and, as a fragment of human work, it is being pushed into technological gaps. This process is linked to a more rigid structuration of the remaining fragments of work according to the requirements of the economy of time (time integration) which goes far beyond the effects already created by Taylorist principles of work organisation. The stricter timing and the integration of these fragments of work are apparent in the increasing standardisation of simpler work, in the subsequent loss of control and the reduction of minimal rooms for manoeuvre over an individual's organisation of work, and in the transfer of the remaining functions (such as controlling, supervising and information processing) to the machinery. The degree of structuration of this fragmented work under the system of the economy of time seems to vary with the proximity of jobs to the automated system. Our research suggests that there are three groups of unskilled repetitive work:

- There are those repetitive tasks in the remaining niches of manufacture which have not yet been automated, such as repetitive jobs in the preparation of electronic components, in soldering, in assembling units, and in testing work in electronic production in the electrical industry which is carried out mechanically and visually.

- Assembly work in the production of cars which has not yet been automated (CO2-welding and other simpler techniques in car-body production) is also unskilled and repetitive. The remaining group of repetitive, but still relatively complex tasks, is at present only marginally affected by the application of computerised technologies and thus has not changed in quality. The development of flexible automation techniques, however, will probably lead to a substitution of this remaining human labour by automated sub-systems in the near future.

- Some repetitive tasks also remain at semi-automated stations, such as manual insertion of printed circuit boards and semi-automatic wiring in electronic production. The technology applied largely determines the content, organisation and time structure of these jobs and they are more or less manual residues from computer-based aggregates or systems.

In addition, in automated systems there emerge unskilled jobs and operating tasks, like simple inserting and assembling jobs, e.g., on welding lines on the fully automated wiring machines and on the automatic test equipment for components. Tasks like inserting can generally be automated by installing automatic loading and feeding magazines. That this has not happened until now, is not due to a lack of technically advanced solutions but to considerations like economic efficiency and return on capital employed.

Tasks like feeding and unloading simple NC-machines and machine systems, which arise under conditions of an increased division of labour, also belong to this category. Generally speaking this involves not very complex skills which can be acquired by "everyman" ("Jedermannstaetigkeiten") and which fill in the missing automatic link manually; but these tasks will gradually be thrust aside by the automation of handling work and intermediate transport.

Thus the traditional type of repetitive work in a Taylorised division of labour is being automated with the growing standardisation in its content and timing and the decreasing complexity of its work operations. At the same time those features of repetitive jobs which are typical of unpleasant, inhuman work are being intensified for the tasks which have not been subject to automation and computerisation.

As a result of the application of computerised technologies in the field of production techniques there emerges a largely new type of automation worker on the level of complex, semi-skilled tasks. This seems similar to the type of complex semi-skilled tasks (like monitoring and controlling production equipment) which has long existed in the steel industry and the chemical industry. This new type results from polarisation tendencies in industrial skilled production work as well as in the traditional semi-skilled work. On the one hand, the application of NC-technologies based on the division of labour produces a type of machine operation and inspection which can no longer be categorised as skilled work (with its craft expertise) but remains at the level of complex semi-skilled labour in its concern with carrying out responsible machining operations and inspection of highly complex technologies. On the other hand the application of production lines controlled by process computers produces simple unskilled functions and complex functions like supervising the whole equipment. These newly created jobs on computer-based machines or machine systems are staffed either with semi-skilled workers with traditional skills who are trained on the new job in the shop, or staffed with skilled craftsmen who normally only remain for an introductory phase and are then replaced by semi-skilled labour as soon as the manufacturing process has become a routine procedure. This type of industrial work is above all characterised by a discrepancy between specific requirements and work stress: as long as the production process proceeds trouble-free, work is neither physically nor mentally challenging whereas contingencies strain concentration and perception as well as intellectual and reactive capacity to the utmost.

The type of skilled automation work on NC-technologies under complex conditions of application changes the traditional qualifications of skilled labour. The specific skill of skilled labourers (which is still almost craft expertise) is largely being substituted by machinery. Where NC-technology is applied the remaining mental, planning types of skill - including programming - change in as much that they are required on a more abstract level and can lead to a condensation of mental and dispositional requirements. Similarly the work of technical inspectors (engaged in checking the function of the plug-in board units) becomes condensed and abstract in localising errors when they have to fill the "gaps in the programme" by acting systematically and strategically.

The type of complex abstract mental work is a result of the application of these computerised technologies which are automating routine work in the offices. The transfer of these functions to machinery means on the one hand that the remaining planning functions are concentrated in the hands of very few employees, whereas on the other hand very simple auxiliary tasks are emerging which assimilate to the type of qualifications which can be carried out by anyone. The remaining complex function of planning work is subject to a decisive change, in as much as the automation of routine results in a condensation of intellectual requirements which, at the same time, are becoming more and more abstract.

These developmental tendencies can be found above all in the application of computer-aided programming and design (CAP/CAD) and in the application of EDP-equipment and other computerised technologies in the planning of fabrication and inspection. With this condensation of work towards highly complex mental requirements there emerges a new type of industrial work; and it must be admitted that not very much is known about its hazards and stress factors. First experiences show that the increase of mental requirements cannot always be classified as a higher qualification and "humanisation" of work. Thus it is crucial for this type of industrial work to find a more balanced structure of highly demanding and regenerating phases.

As to the structural features and the developmental tendencies of these four characteristic types of work in fabrication technique, there are signs of a polarisation process which produces new profiles of stress and requirements. The application of computerised technologies interferes deeply with the planning, dispositional, controlling and supervising functions of human work, by incorporating increasingly more fragmented parts of these functions into the machine. This process leads to unskilled, repetitive, fragmented work which is experienced by the individual worker as involving a loss of control over work; and to highly complex, abstract mental work, which condenses the planning parts of formerly undivided work. Between these two types of industrial work emerges a type which is relatively new to the fabrication process. It can be distinguished as complex semi-skilled work (after a short period of training) and will become more important when automation and the integration of the fabrication process become more advanced. The remaining type of skilled work, however, seems to be confined only to a transitional phase and seems to combine complex semi-skilled work and elements of the intellectual-planning type. This type is always in jeopardy of being fragmented and rendered superfluous by further automation or a centralised organisation of work.

Consequences for Bargaining Policy and Trade Union Organisation

How does this affect the relationship between skill requirements and trade union bargaining and organisation policy? Apart from structural unemployment, changes in working conditions and requirements are likely to become the central issues in all future trade union strategies concerned with the organisation of work and the workplace and the protection of skills. Whilst the introduction of the 35 hours working week is widely discussed at present as a concrete step towards reducing unemployment, there is a lack of strategies for work organisation and the protection of skills. The regulations of the Works Constitution Act of 1972 (article 90, 91) certainly exist but they do not give any power to the works council over the organisation of work. The complexity of problems which have arisen mean that these regulations have proved to be insufficient under conditions of an increasing computerisation of production.

With regard to work organisation and skill protection, there are two important aspects which have largely been neglected by the trade unions: they are still using a definition which is dominated by the traditional skill and capacity of the highly skilled craftsman. This notion, however, is out of date and thus unfit for comprehending and defining all the new skill elements. Similarly neglected are the prerequisites for exerting an influence on the organisation of working conditions in the work shops. In this context the trade unions' underlying appreciation of the relationship between technology and work organisation, is likely to be of central importance. The often discussed assumption that the "plasticity" (or flexibility) of computerised technology would increase the room for manoeuvre in matters of work place design (layout) is quite unrealistic. The possibilities for exerting an influence on the work organisation at shop floor level under conditions of computerised production are very limited, if they exist at all, and are confined to ergonomic rather "cosmetic" interventions as to the layout of external work-facilities. Alternative forms of work organisation, such as shop floor programming, which are introduced in the application of CNC-machines in machine building, only seem to be exceptions. In practice they serve to legitimate changes within a general framework of entrepreneurial technology strategies which aim at a long term displacement of skilled labour in the manufacturing process. The "on-the-spot-programming" ("Vor-Ort-Programmierung") of CNC-machines requires a certain layout of machine control which is rejected by the companies for reasons of profitability as well as for reasons of securing power and domination in the work shop. The successive introduction of directly numerical controlled systems (DNC) for a central management of programmes indicates the future relocation of programming from the shop floor to central programming departments. Corresponding to the degree of incorporation of work organisation measures into the development and design of technology, the possibilities of an alternative application of technology in the plants are largely being restricted. What matters in trade union strategies with regard to workplace layout and skill protection is, therefore, to exert influence on the technology policy and thus on the development and design of technologies. This necessitates a co-ordination of trade union efforts not only on the workshop level but on the level of company and corporation. In this context an internationally co-ordinated strategy among different national trade unions to exert influence on the technology policy of capital should become a matter of urgency.

The computerisation of production and work processes deeply changes both the conditions of human labour at workshop level and the character of work itself as the production of surplus value. While the central points of Taylorism and Fordism were to intensify individual work and to utilise workers' co-operation potential, it is now - in the most technologically advanced production - the "collective worker" ("Gesamtarbeiter"), who is the focus of all entrepreneurial strategies for the production of surplus value. The efficiency policy at shop level does not hinge upon speed as a central principle, but on the timely and appropriate actions of the workforce. To be able to react quickly and efficiently in the case of a contingency and thus avoid possible damage and stoppage to capital intensive production equipment, is the central principle for the employment of human labour in a computerised workshop.

The application of computerised technology also changes the different elements of skill. The increasingly abstract function of work and the requirements of the production equipment can be observed most markedly in the increase of qualifications which can no longer be described in terms of traditional skill. As a corollary there is a tendency for psychic strain to replace physical burdens. This can be explained by the intensification of new work requirements (e.g. attentiveness, concentration, responsiveness, responsibility etc.).

The central principle of the entrepreneurial strategies to maximise efficiency is the increasing demand for a flexible employment of the workforce. The flexible employment of labour, as the key to strategies at workshop level, as well as the changes in work requirements and working conditions, are breaking the constraints of traditional systems of efficiency rating. In this process the trade unions are obliged to develop new systems of rating, grading and pay which are compatible with the utilisation of human labour in technically advanced processes of production.

The erosion in the field of skilled labour, on the one hand, and the development of plant-specific skill requirements in highly advanced technical processes of production, on the other, have a lasting effect on the development of the job market. Apart from the markets for specific skills and the general market for unspecific unskilled labour the technical development especially in large companies leads to sectional plant-specific skills. Due to this new segmentation of the job market there emerges a newly composed regular core-workforce

("Stammbelegschaften") within the companies, with qualifications which are tailored for the technical system of the plant. In this context the loyalty of the employees towards their company and identification with the economic targets of the company plays an important role in the relationship of company and regular (core)-workforce. Whereas under conditions of conventional production techniques, the entrepreneur was dependant on the workforce for its specialised skill and knowledge about the production process, in conditions of an increasing computerisation of production he is dependant on the loyalty of this core-workforce. The identification with the targets of the company thus becomes the basis of a new arrangement between management and workforce. Apart from this loss of skilled labour which leads to a weakening of the trade union's power in the plant - because traditionally the skilled workers have been the most important basis for recruiting members - the development of a core-workforce in the plant becomes an additional challenge for the trade unions. There is no doubt that the permanent core-workforce needs the trade unions because their interests further depend on the overall situation on the labour market and thus require a general protection: but the trade unions are more dependant on these core-workforces because they constitute their basis of power in the companies - in a situation which is characterised by an enterprise-centred segmentation of the labour market (Deutschmann/Schmiede 1983, p.42). Altogether the changes of the skill structures in the companies - being a corollary of the development of technology - are confronting the trade unions with the need to rethink their traditional collective bargaining and organisation policy and to adapt to the new conditions on the micro- and macroeconomic level if they want to maintain their function as organisations of interest representation in the future.

REFERENCES

Benz-Overhage, K., E. Brumlop, Th.v. Freyberg, and Z. Papadimitriou, 1982: Neue Technologien und alternative Arbeitsgestaltung - Auswirkungen des Computereinsatzes in der industriellen Produktion, Frankfurt/New York.

Benz-Overhage, K., G. Brandt, and Z. Papadimitriou, 1982: Computertechnolien im industriellen Arbeitsprozess, in: Schmidt, G. et al. (eds.)

Materialien zur Industriesoziologie, Sonderheft 24/1982 der Koelner Zeitschrift fuer Soziologie und Sozialpsychologie, Opladen, pp. 84-104.

Brandt, G., B. Kuendig, Z. Papadimitriou, J. Thomas, 1978: Computer und Arbeitsprozess - Eine arbeitssoziologische Untersuchung der Auswirkungen des Computereinsatzes in ausgewaehlten Betriebsabteilungen der Stahlindustrie und des Bankgewerbes, Frankfurt/New York.

Brumlop, E., W. Rosenbaum, 1979: Humanisierung der Arbeitsbedingungen durch gewerkschaftliche Tarifpolitik, in: Bergmann, J. (ed.), Beitraege zur Soziologie der Gewerkschaften, Frankfurt, pp. 264-97.

Deutschmann, Ch., R. Schmiede, 1983: Lohnentwicklung in der Bundesrepublik 1960-1978 - Wirtschaftliche und soziale Bestimmungsgruende, Frankfurt/New York.

Kern, H. 1979: Kampf um Arbeitsbedingungen - Materialien zur "Humanisierung der Arbeit", Frankfurt.

Lutz, B. 1978: Wirtschaftliche Entwicklung, betriebliche Interessen und Arbeitsmarkt-segmentation, paper, Muenchen.

Lutz, B. 1982: Kapitalismus ohne Reservearmee? Zum Zusammenhang von Wirtschaftsentwicklung und Arbeitssegmentation in der europaischen Nachkriegszeit, in: Schmidt, G. et al. (eds.), Materialien zur Industriesoziologie, Sonderheft 24/1982 der Koelner Zeitschrift fuer Soziologie und Sozialpsychologie, Opladen.

Ohm, Ch., S. Wenk, 1981: Integration von Technologie- und Qualifikationspolitik als Problem gewerkschaftlicher Strategie, in: Jansen, D. et al. (eds.), Technischer und sozialer Wandel - Eine Herausforderung an die Sozialwissenschaften, Koenigstein/Ts., pp. 67-85.

Vetter, H.O. (ed.) 1974: Humanisierung der Arbeit als gesellschaftspolitische und gewerkschaftliche Aufgabe. Protokoll der Konferenz des DGB vom 16. und 17. Mai 1974 in Muenchen, Frankfurt.

Chapter Three

TECHNOLOGICAL CHANGE, LABOUR MARKET, AND TRADE UNION POLICY

Gerhard Brandt

The relationship of technological change, labour market structure, and trade union policy has long been neglected in social and economic research. This neglect is certainly related to the division of labour between industrial relations research, industrial sociology, and economics. For, even when individual aspects of this relationship are dealt with by individual disciplines and their prerequisites and consequences at other "levels" are indicated, the relationship as a whole remains unexamined. But this neglect is also materially grounded in the fact that, under the impact of continuing prosperity, official trade union policy has long neglected the connections between production and politics. Instead it limited itself to ensuring that employees shared in the benefits of increased productivity.

The significance of this issue has become quite obvious today in both academic and political circles owing to the continuing crisis in economic performance and employment. This raises three key questions: what consequences do technological change and its repercussions on labour market structure have for trade union policy? Secondly, how can unions meet these changes? And, thirdly, what does this imply for the relationship between union policies and state policies? In addressing the first two questions in a preliminary manner, we will start out from West German experience. Here the structure of production - as mediated above all by the dominant labour market structure - exercises a strong influence over trade union policy. We will then examine the cases of the United States, Great Britain, and Italy to see if, and to what extent, the link between production and trade union policy differs from the West German pattern.

West Germany: a Case of Neo-Fordism?

Our understanding of the West German case has been considerably revised in the course of our research. Initially we interpreted union demands in the areas of wages, work time, working conditions, and employment in terms of the role of unions as "intermediary organisations" which mediate among the interests of their members, the constraints of the capitalist system, and the political framework established by the state (Bergmann et al., 1975; on the background to our research, see also Billerbeck et al., 1982; Brandt et al., 1982; and Deutschmann and Schmiede, 1983). In this context unions serve as just one of the negotiating agents in the dual system of collective bargaining (Tarifautonomie) in the Federal Republic. For trade unions reserve for themselves negotiations over the norms which bind particular economic sectors regarding wages, the conditions of wage payment, the duration and disposition of labour time, and more general working conditions. The concrete regulation of wage payments and working conditions, on the other hand, is left to the Works Councils. These were established by law and act as bargaining agents on the enterprise level. We believed that the remarkably co-operative character of trade union policy in West Germany would last as long as, firstly, capital's scope for concessions enabled employees to participate in the growing national product in the neutral form of money income and leisure time, within the limits of increased productivity, and as long as, secondly, immediate working conditions at plant level could be protected and improved in the face of the rationalisation strategies pursued by capital. As the scope for economic concessions diminished and as forced rationalisation of production developed in the mid-70s, we believed that trade unions would turn from their co-operative policy to a more conflictual, "politicised" approach.

We anticipated a twofold shift in union policy. Firstly, with the decreased opportunities for buying off qualitative demands through quantitative wage concessions, trade union policy could be expected to shift from wage policy to a concern with working time, working conditions, and employment. We defined this as a "horizontal" displacement. Secondly, owing to the limited capacity of the Works Councils to solve problems at plant level, we also expected a "vertical" displacement from the enterprise to the sectoral level of collective bargaining.

These expectations were largely disappointed. The co-operative policy grounded in the apparent stability of the dual system continued to operate with at most a limited thematic shift from quantitative to qualitative demands. Two factors seem to have been at work. We had underestimated the problem-solving capacity of trade union and enterprise bargaining (cf. Streeck, 1979) and we had taken too literally the unions' claim to represent the class interests of all workers and thereby neglected the previous role of labour market segmentation in the formation of union policy (cf. Lutz, 1982). In the latter respect some employee groups are more highly organised and this is reflected in their disproportionate representation in membership figures and, what is more, sections with privileged market positions gain even more influence in the representational system in so far as they are usually over-represented in the unions' decision-making bodies. Here they can influence the formation of trade union policy and thereby reinforce the process of selective recruitment. Although one could not say that the West German industrial unions represent the interests of the old and the new labour aristocracies or are disguised craft or enterprise unions, their undoubted role in generalising the diverse interests of particular groups of employees remains limited to the relatively abstract themes of general wage increases and the regulation of labour time. The policy of the trade unions actually emerges from extremely selective forms of processing and consideration of employee interests which, beneath the level of "abstract generalisation", reproduce the unequal distribution of material employment and life chances. Moreover, in contrast to their ideology and programme, this aids and abets the transfer of the effects of crisis to already disadvantaged employment sections. In this way trade union policy corresponds exactly to the mechanisms of labour market segmentation and reproduces the latter at the level of interest representation through the selective processing and distribution of instability and uncertainty (cf. Berger and Piore 1980). Both the stability of the industrial relations system and the continuity of trade union co-operation can be decisively explained by their common basis in labour market segmentation. They also follow the same logic as the latter. It remains to be seen whether this functional relationship persists or will be questioned as the employment crisis invades the different segments of the labour market.

This correspondence thesis has been developed in a stronger version which relates labour market structure, strategies of capital, and the production structure (cf. especially Schmiede, 1983; Deutschmann and Schmiede, 1983). This stronger thesis suggests that the general process of labour market segmentation and, in particular, the formation of demand-orientated and internal labour markets correspond to secular changes in the production structure of capitalist industrial societies. In turn these changes are said to express a new mode of capitalist social integration. This is supposedly based upon the progressive mechanisation and automation of industrial production, which replaces human labour by machines and, in Marxist terminology, increases the organic composition of capital. In operational terms this tendency can be observed in the relative increase of fixed capital, in increased capital intensity and/or a higher capital coefficient and, in view of the sheer expansion of technical production units in mass and large batch production, in a concentration of capital. According to the thesis, the mechanisation and automation of production (which was initially carried out more randomly) reaches a critical threshold which forces a reorganisation of the production process. Whereas this reorganisation at first spread in the form of technical rationalisation and, in the science-based industries, as scientification, it was later extended to human labour in the form of rationalisation of labour organisation (cf. in detail Schmiede and Schudlich, 1976).

Since the most important aspects of this process are well known we need only recall the three main changes which are reflected in the labour market structure. Firstly, there is the breakdown of traditional craft skills and their replacement by specialised kinds of semi-skilled work which are enterprise-specific and can be learned, generally through on-the-job training, in a relatively short time. Owing to the critical role of the "residual functions" of human labour in automated production processes, however, this semi-skilled work requires a high degree of work motivation and flexibility. Secondly, there is the integration of human labour into the mechanised and automated production process with the aid of practices of labour management which aim to synchronise the performance of machines with those of human labour power. And, thirdly, there is the formation of a permanent core of employees as the key to an employment policy which aims to ensure that a complex system of connected plants can function independently of short-term production fluctuations. Taken together, these measures clearly lead to the formation of an internal labour market which offers the members of the permanent staff relatively high job security but also ties them quite closely to the enterprise. Indeed, owing to the enterprise-specific nature of their qualifications and experiences, a change of firms would appear to have little chance of success as a means of occupational and socio-economic advancement. In theoretical terms, these measures can be understood as instances of a new mode of capitalist integration which no longer operates through the exchange of commodities in the market as the objectification of human labour time but occurs directly in the production process itself on the basis of a unified measure of abstract time (Sohn-Rethel, 1978; Schmiede, 1983) and via the synchronisation of machines and human labour. Seen in this way, policies concerned with qualification, labour management, and employment and, as their result, the labour market structure of the enterprise appear as the forms of expression of an economy of time that is materialised in the production structures of large integrated enterprises. Accordingly they stand in contrast to the traditional market economy.

The proponents of the stronger version of the correspondence thesis are not arguing that the reorganisation of the production process and the concomitant creation of internal labour markets are established out of sheer necessity. Instead they are mediated by the strategies pursued by capital. Nor do they deny that these strategies are the result of management decisions which reflect different decision-making situations from case to case. They merely maintain that management is obliged to ensure the existence of the enterprise in the face of the uncertainties of technical and economic development and that, given complex and highly integrated production units, policy measures including qualification, control of the labour force and employment hold the best promise of achieving this goal on the basis of time economy. According to this thesis the managements of large industrial enterprises also serve as a mediating instance in the industrial relations system and must take account of the logic of the existing economic system with due regard to variable environments. It is only in this way that the structure of the labour market (and the trade union policy mediated through it) appears as the "reflex" of the changing industrial structure of production and, more basically, of the real subsumption of labour under capital which is expressed therein.

Contrasting Evidence: the United States, Great Britain, Italy

The West German model developed by the Frankfurt project group was influenced by the labour market segmentation discussion which began in the United States in the early '70s and then spread to Western Europe. Our approach shares several fundamental assumptions with the "dualists" as well as the "radicals" (cf. Rubery, 1978). Like the dualists, we assume that the segmentation of the labour market is based upon developmental tendencies of production techniques and/or production structures which separate primary from secondary, and internal from external, markets. Like the radicals, albeit without voluntaristic connotations, we assume that the segmentation of the labour market also contributes to the division of the collective body of employees so that they can be more effectively ruled. But this means that the model is subject to the same critical objections which have recently been addressed to both the dualist and radical versions of the segmentation thesis (cf. Rubery, 1978; Koehler, 1981; Sabel, 1982).

If one follows the empirically based reasoning of Sabel, the West German case, as we have presented it, corresponds to "neo-Fordism". But there is nothing inevitable about this particular relationship of production and politics and, in Sabel's view, it actually constitutes a pathological deviation (Mueller, Roedel, Sabel, Stille, Vogt, 1978, document Sabel's affinity to the Starnberg School just as does the first chapter of his latest publication). A theoretical critique is also implied in these empirical objections. For the segmentation thesis in both the dualist and radical versions reduces the predominant structure of the labour market and its correlative trade union policy to supposedly elementary conditions of production. In this respect it fails to recognise that "the present form of technological dualism ... was and is not a technical necessity." For "what is a fixed and what a variable cost ... depends on political and institutional factors that cannot be deduced simply from examination of the technology of production" (Sabel, 1982, p. 45).

Sabel's critique of the technological determinism of the neo-Fordist model exemplifies an antiobjectivistic counter-current within discussions about the development of the organised labour process under capitalism and the relationship between politics and production (cf. Burawoy, 1978). It also has clear implications for our model as well. The arguments presented here provide reason enough to re-assess the decisive claims on which this model rests. It would seem best to begin with the weaker version of our thesis, which deals only with the relationship of trade union policy and labour market structure; and then to consider the stronger version, which extends to the relationship between the labour market structure and the production structure. Thus we first present a very preliminary, internationally comparative survey of different forms of the relationship of trade union policy and labour market structure; and then we briefly consider the problematic nature of dualism as it appears in more recent publications. Our remarks are largely empirical; more fundamental theoretical objections must await another occasion.

If one confines oneself initially to the first, weaker version of our thesis, presupposing the cleavage of the labour market into primary and secondary and internal and external segments, then an obvious objection is that trade union policy is not unequivocally determined by labour market structure. Instead, without doubting the latter's relevance, trade union policy might be determined by the programme of trade unions and, ultimately, by the interests of their membership. Although employee interests are certainly influenced by the labour market structure and are also subject to selective processing by the trade union organisational structure, they are not entirely defined by such influences. According to Sabel, employee interests are based on "world views", which represent in turn the product of experiences of struggle and learning. Thus they present an autonomous moment which varies from epoch to epoch and from country to country even if conditions are otherwise the same. It is these world views which guide the unions in their confrontations with the given structural determinants (Sabel, 1982, p. 14 ff. and passim). Whether and how the policy of the trade unions varies in relation to labour market structures from one country to another and what this implies for the situation of the workers (and for the structuring of the labour market), can be seen from several national cases.

The case of the USA provides a good starting point for an international comparative survey since it is quite well documented (cf. especially Edwards, 1979; Koehler, 1981; Koehler and Sengenberger, 1983). The available literature suggests that the initial inter-company splitting of the labour market in the American automobile industry was the "result of an economic separation of the branch into relatively stable and relatively unstable productions" which had been planned quite early and was accomplished without the active participation of the United Automobile Workers (UAW) founded in 1936 (Koehler and Sengenberger, 1983, p. 415 f.). In this way the companies obviously attempted to use the surplus supply of labour power that continued into the late 30s for a flexible employment policy in this newly delimited secondary sector. In the primary sector the splitting was followed by the elimination of "technical" control over the labour force which had been practised up until then, based predominantly on technological and labour-organisational constraints; this was now replaced by "bureaucratic" control orientated to formal rules (cf. Edwards, 1979, p. 130 ff.). This step already implied a repeal of "despotic" practices and was connected with a rationalisation of company authority in the primary sector (cf., besides Edwards, Burawoy, 1979, p. 104 ff.). This process was carried on after the war by company agreements with the UAW, which was bureaucratically organised and firmly settled by now, with its membership base largely among the semi-skilled and unskilled labourers. The UAW limited itself to the regulation of working conditions and wage agreements and the regulation of personnel movements according to the standards of a seniority system. Thus the prerogative of management to make production decisions and related personnel decisions was not questioned. None the less these decisions and their implementation were related to binding and agreed-upon rules, adherence to which led to the decline of turnover and meant a considerable increase of inner-company mobility along prescribed routes as well as a distinct increase of job security (cf. Burawoy, 1979, p. 97; Edwards, 1979, p. 158 f.). One can agree that "tailor-made seniority systems", as practiced in the organisational realm of the UAW and common in other branches of American industry, do "correspond to company interests in their basic structures, but, at the same time, employee interests and trade union policy exercise a significant influence on the structuring of the company labour market" (Koehler, 1981, p. 349).

A policy, such as that of the UAW, which aims to subject the enterprise labour market to some kind of control, even if only formally, can apparently transform the traditional "despotic" forms of enterprise authority into "hegemonic" forms involving the participation of workers organised in unions (cf. Burawoy, 1979, p. 110). The frequent objection that such a policy only benefits the workers in the primary sector and encourages the fragmentation of the working class falls short of the mark. For the regulation of seniority extends far beyond the existing permanent labour force; and the policy of these unions has helped to transfer a considerable portion of semi-skilled and unskilled workers into the "subordinate primary market" (Edwards, 1970, p. 170 f.). And, although the radicals in particular have insisted that the policy of American trade unions, directed at the regulation of the segmented labour market, aids and abets class divisions, even they admit that this regulation also involves a fundamental challenge to enterprise authority and thereby opens up perspectives for redressing the balance of class relations and producing a substantial democratisation (cf. Edwards, 1979, p. 200 ff.). More recent investigations point out, of course, that the traditional collective bargaining system in the mass production industries of the U.S.A. is being progressively eroded under the influence of continuing mass unemployment and as a result of fundamental changes in the production and economic structures at the cost of large-scale industrial production (cf. Piore, 1983). Thus the system of formalised control that has been supported by the unions will probably also be affected.

The case of <u>Great Britain</u> is less well documented (exceptions to this would be the study of Rubery 1978 and that of Friedman 1977). Nevertheless one can conclude that the division of the labour market is also a central problem of industrial relations in Great Britain and that it is more likely to assume the form, at least in the automobile industry, of an inter-enterprise cleavage into internal and external markets at the international and national level. In this respect it is similar to the United States rather than Germany with its internal division into a permanent and a marginal labour force (cf. Friedman, 1977, p. 240 ff.). However, in contrast to the United States, the segmentation of the labour market in Great Britain cannot be so clearly traced to management strategies to which the trade unions have merely passively, and with a delay of time, reacted. Instead it is probably based on the interaction between management and organised workers, if not almost entirely traceable to the actions of the latter. "Trade unions have probably played a more active role in the formation of structured labour markets in the UK than in the US" (Rubery, 1978, p. 32, and even more emphatically Burawoy, 1979, p. 189, who states that the internal labour market in Great Britain, in contrast with that of the United States, was "... largely organised by the union rather than by management"). One reason for this was the lack of a larger reserve force of pre-industrial labour which would have allowed management to develop a strategy of internal external labour market division. In addition the early unionisation according to occupational groups enabled the workers to offset the effects of the mechanisation and rationalisation drives between the Wars and in the post-war period through organised action. For quite a while, traditional craftsmen's groups succeeded in maintaining their traditional status when faced with de-skilling tendencies. At the same time, ascending groups of semi-skilled and unskilled workers, such as those in the automobile industry, attained recognition as negotiating parties by organising alliances within the framework of established or new organisations. Thus they were able to enjoy the privileges that membership of an internal labour market can afford (Zeitlin, 1980). "De-skilling of the labour force in the UK thus often led to an extension of organisation to semi-skilled and unskilled workers and some reduction in differentials, whilst a structured, sheltered labour market, based on control of entry, was still maintained" (Rubery, 1978, p. 31). Upon this basis it was apparently possible to make "job ladders, skill demarcations and the pace of work" the subject of binding agreements (ibid., p. 29) and to secure protection from the influences of the external market.

This growing demarcation of internal from external markets was supported by traditional organisations as well as by new and competing ones (cf. e.g., Brandt, 1975, on this development in the steel industry around the turn of the century) and/or through shop stewards activities. But these did not go beyond pragmatic alliances and could not lead to a united front or a coherent alternative to management strategies. Labour market segmentation in Great Britain appears much more emphatically than in the United States as the resultant of power relations between management and trade union or shop-level organisations. In turn this gives workers and their organisations, e.g., within the framework of the closed shop, the chance of largely controlling working conditions as well as internal and external mobility. In the absence of detailed institutionalised regulations such as those familiar in the practice of the UAW and the American steel workers, however, the chances of such control could be undermined by adverse economic and political circumstances. Indeed, one can identify apparently quite successful attempts by management, as in the chemical industry in recent years, to organise internal labour markets according to the pattern of hegemonic, if not despotic control (cf. Nichols and Beynon, 1977; Nichols and Armstrong, 1976, and the reference by Burawoy 1979, p. 189).

The most complex and contradictory, yet highly instructive and even exciting, case dealt with here is that of Italy. It has the added advantage of being most extensively documented in everyday political discussions as well as in the social science literature (if, however, primarily in Italian), and it has been thoroughly analysed by authorities as well as critics of dualist segmentation theory (cf. especially Berger and Piore, 1980, and Sabel, 1982). Of decisive significance for understanding Italian trade union policy is, in the opinion of every author, the role of the traditional sector. If one shares the argument of Berger and Piore, this sector comprises an integral element of existing industrial society and is not simply a relic of a pre-industrial mode of production. It consists of small and medium-sized agricultural and manufacturing enterprises that are usually operated by their owners. This contrasts with the USA, Britain and, one might add, with Germany, but holds for Italy and, with several modifications, for France.

The importance of this traditional sector after the war for the large industrial combines, located primarily in northern Italy, was the possibility of being able to meet the growing demand for workers in a flexible fashion, without being subject to the pressure of a labour force having experience in industry and being well organised as happened in Great Britain. The labour market strategy of Italian industry, which lasted well into the 60s, appears to have consisted of an internal splitting of the enterprise labour market into permanent and marginal groupings. This was made even easier by the weakness of the Italian trade unions, which were split into associations organised along the lines of party politics (Berger and Piore, 1980; Sabel, 1982). A complex relationship of learning experiences and struggles, in which militant political cadres and students as well as masses of semi-skilled and unskilled workers participated, is responsible for the alliance formed in the course of the 60s and which was able to triumph over the entrepreneurs under the union leadership in the "autunno caldo" of 1969. The success of these politics based on the tradition of "operaism" was based on a re-organisation of enterprise bargaining and the unions' acquiring extensive control over plant-level working conditions and employment policy (cf. in detail Brandini, 1978; Regalia _et al_., 1978; Sabel, 1982). One could even say that the division of the internal labour market was eliminated and the kind of flexible employment strategies that had previously been exercised at the level of the firm were undermined.

The solution that management found after it had been robbed of its internal freedom of action consisted of displacing portions of the production volume into the manufacturing part of the secondary sector. The elimination of the internal labour market division that was accomplished, albeit unintentionally, by the collective action of the unions entailed a reactivation of the traditional sector, involving an intensification of the external splitting of the labour market. While discrimination among the differing forms as of industrial labour was overcome in the industrial centres, it was frequently reproduced in a much more drastic form in a secondary sector which had experienced little union organisation (cf. Berger and Piore, 1980). Of course, if one adheres to Sabel's interpretation, even the extreme external segmentation itself contains a progressive, even revolutionary potential, since the small-firm forms of production as carried on along the lines of a cottage system by politically sensitised groups represent the elements of an alternative economy (cf. Sabel, 1982, p. 220). The Italian model would amount, accordingly, to an alternative to neo-Fordism as practiced in the United States and in the Federal Republic. For, although the breakthroughs in the industrial centres have largely been rescinded under the pressure of the employment crisis, still other breakthroughs constitute an alternative due to the unintended consequences at the periphery.

Production and Politics

Our comparative survey permits a clear judgement on the weaker version of the neo-Fordism thesis. Trade union policy and, in a wider perspective, employee actions cannot be reduced to mere reflexes of the structure of the labour market. Nor should they be seen as dependent variables of it. Instead they represent quite independent forces within certain limits. Thus, rather than using categories such as "reflex", "dependent" and "independent variables" connoting a one-sided relationship between trade union policy and labour market structure, it would seem more appropriate to speak of "challenge" and "response" even when the structure of the labour market is largely the result of the strategies pursued by capital. Thus one can grasp trade union policy and employee actions as politically motivated responses to the challenges of a given labour market structure. The nature of these responses and their impact on these challenges would appear to depend decisively upon the developmental stage and "timing" of trade union organisation. A particularly decisive question is whether the organisation of workers preceded the structuring of the labour market or followed it (cf. Burawoy 1978, p. 189).

Our suggested approach could still be reconciled with the assumption that the policy of the trade unions represents a purely extrinsic response to the given and persisting structure of the labour market. This might be supported by the case of the American automobile and steel industries. But comparative analysis enables us to recognise that the political activity of workers at the level of the enterprise or sector is an active factor (or, at least, can be). Indeed, as the British and Italian examples clearly demonstrate, it has a modifying and formative effect on the structure of the labour market (cf. especially Sabel, 1982, p. 70). Thus trade union policy and labour market structure should be seen as involved in a reflexive, feedback relationship in which they interact with capital's own strategies. This is also admitted by defenders of the dualistic theory of labour market segmentation whose more recent works recognise the efficacy of political and institutional factors in the "actual structuring of the (labour) market" and thereby expand the original framework of their theory (cf. Berger and Piore, 1980, p. 72). Thus Piore has suggested a twofold concept whereby the instability and uncertainty which result from labour market segmentation are grasped, on the one hand, as a "common problem of industrial society" allowing quite different solutions within certain objective conditions of social action. On the other hand, segmentation is seen as a variable that is set in the process of the advancing division of labour and which "affects" this process (ibid., p. 79 f.).

Our findings also have implications for the stronger version of our thesis. If it is true that the structure of the labour market is one of the parameters of trade union policy, if this structure varies from one country to another and from one branch to another, and if it is influenced by this policy and/or the struggles between management and the unions, then it cannot be considered a "reflex" of the structure of production. The labour market structure as the product of past or present social struggles is independent of the structure of production and affects, within certain limits, the formation of the structure of production (cf. especially Sabel 1982, pp. 70, 126, 176, 230 f.). In short, in talking about connections between labour market structure and the structure of production, we are dealing with reflexive relationships and second order reactions, which are embedded in the all-embracing relations of political action and the structure of production. In this sense the labour market structure functions as a level or instance of mediation between politics and production. An even more significant revision of the objectivistic assumptions of dualistic segmentation theory occurred when recent studies examined the structural discontinuities in industrial societies. This distances such studies from conventional industrialisation and modernisation theories and their implicit logic of technical and organisational determinism. If structural discontinuities assert themselves in developed industrial societies in the form of a self-perpetuating traditional sector and, one might add, quasi-corporate forms of trade union organisation and if these discontinuities, as can be seen in Italy and France, tend to become effective as a second economy in the structure of production, then the latter can no longer be understood as an independent variable in relation to the structure of the labour market and in relation to union action. Instead, one must develop a theory of the dual market that is technologically and politically founded. This is already evident in the more recent works of Berger and Piore.

These largely implicit and unsystematic findings from industrial relations and labour market research have been supplemented and confirmed by technology research undertaken by social scientists and historians in the United States. In opposition to technological determinism these researchers conceive technological development as a social project. They are often inspired by the Marxian and critical theory traditions and, in contrast to many participants in the present technology discussion, are not content with speculative and conceptual reflection, but instead follow a strict empirical-historical research programme (cf. especially Noble 1977 and 1978). Particularly significant for our question is the twofold thesis that the production technologies implemented in the fields of information and communication as well as industrial production are not the reflex of a unique logic of technological development but derive instead from social choices. These choices are made at the level of applied research and development in the light of economic requirements. But these technologies also reflect struggles and conflicts at the sectoral and even more at the enterprise level over the conditions and consequences of their implementation. It is here, and only here, that labour has the chance "to register its choices in the process of technological development" (Noble, 1978, p. 320). Choices about the introduction or not of a production technology, about its modification or the conditions of its implementation present themselves to workers as choices over the control of direct working conditions and, to this extent, are also decisions about the (internal) labour market structure. In so far as workers take advantage of this chance, the structure of production, as mediated by the labour market structure, proves to be a parameter, and not simply a determinant, of workers' actions.

Although originating in diverse theory and research traditions, all these studies seem to share some uneasiness about the one-dimensional development paradigm found both in conventional industrialisation and modernisation theories and in traditional Marxist theory. Likewise, they all try to overcome the weaknesses of this paradigm through a more complex and multi-dimensional approach. The reference point and point of attack of the Marxist orientated studies is the extremely influential work of H. Braverman (1974). The latter is criticised for having presented the capitalist production process as simply the expression of objectively effective mechanisms of system integration and thereby neglecting all of the subjective aspects and moments of resistance based in class antagonism (cf. especially Burawoy 1978, but also Noble 1977 and Sabel 1982). Such an analysis, according to these arguments, not only promotes political resignation (a criticism that is not very compelling theoretically), but can also be criticised for presenting a simplified and distorted picture of social reality. Works such as those of Burawoy, Sabel and Nobel try to consider the subjective and political moments neglected by Braverman and his followers. They often point to precedents set by E.P. Thompson and E.D. Genovese and attempt to reconstruct the history of the capitalist production process as the history of class struggle (cf. Burawoy, 1978). Thus neo-Fordism, which was supposedly based exclusively on objective constraints, is now seen as an improbable and pathological marginal case which has been purged of the tensions and disputes to which the everyday world of the shop floor gives rise. In the light of German experience, it appears dubious even to us that this form of economically coerced adjustment can be practised over an extended time period.

The Thesis Reconsidered

The preceding analyses suggest that our thesis needs considerable revision but need not be entirely abandoned. It should be clear that the relationship of trade union policy, labour market, and the structure of production cannot be conceived in terms of a one-sided dependence or in terms of the base-superstructure scheme. It seems more appropriate to understand these factors as instances of a dual hierarchy which becomes effective, on the one hand, in controlling activities (hierarchy of controlling factors) and, on the other, in restrictive conditions (hierarchy of conditioning factors) (cf. Parsons, 1966, p. 28). In these terms union policy serves to control activities in the labour market and the labour market in turn helps to control the structure of production. At the same time, the structure of production operates as restrictive condition in relation to the labour market and this, in turn, conditions the policy of the trade unions. Secondary analyses may reveal whether this reformulation of our model leads to different results (see the figure on the next page).

This new model should be demarcated from rival theoretical positions in at least two respects. We deal first with communication theory. Trade union politics cannot be equated with normative world views and communicative action; nor can the labour market and the structure of production be equated with the dimension of labour and instrumental action. For one must proceed from the fact that specific historical processes of capitalist social integration influence all of these factors and ensure a progressive abstraction from all qualitative properties. Our investigations consider this tendency in terms of a "double selectivity". This concept identifies how the qualitative issues of trade union politics are reduced to quantitative issues and the latter, in turn, are adjusted to the functional prerequisites of the existing economic system. We firmly believe that this concept and its underlying assumptions have been, and will continue to be, confirmed in our investigations.

Our approach also differs from purely formal cybernetic theories to the extent that the latter presume a complete symmetry or equivalence of control and conditioning hierarchies. Although a considerable margin can be demonstrated for the control functions of trade union politics during phases of prosperity, the continuing employment crisis reveals that this margin is drastically reduced through increasing restrictions at the level of the labour market and the structure of production. We are convinced that the decrease in control chances exercised by employees' representation on the shop level in Great Britain and Italy - which is not denied by authors such as Sabel - can best be explained in this way. Just as the fact that shop level and trade union representatives are forced to conform to the constraints of capital valorisation. The current erosion of the system of industrial relations in all the countries dealt with here reveals the increasing

Figure 1. Trade Union Policy, Labour Market and Structure of Production within the Context of Cybernetic Relations and Modes of Social Integration

	<u>Structural Levels</u>		<u>Modes of Social Integration</u>
	Trade union Policy		Double Selectivity of Union Structure
Hierarchy of Conditioning Factors	Labour Market	Hierarchy of Controlling Factors	Market Economy
	Structure of Production		Time Economy
		Cybernetic Relations (Control vs. Restrictions)	
		Modes of Social Integration (Advancing "Abstractification")	

impact of economic constraints under conditions of mass unemployment.

Several clues suggest that neo-Fordism of the type prevailing in the Federal Republic was largely based on the "overadjustment" practised by West German unions from the early 50s (cf. Brandt et al., 1982). But it would be dangerous to disqualify it as a "deviant" or "pathological" pattern since this pattern has also emerged elsewhere under the impact of the deepening employment crisis. A marginal case under the auspices of prosperity, it might become normal under conditions of mass unemployment and structural change. How this happens, what specific course the development takes, and whether this is really the final word, is a question that must await further research and, indeed, political practice.

REFERENCES

Berger, S., Piore, M.J. 1980: Dualism and Discontinuity in Industrial Societies, Cambridge: Cambridge University Press.

Bergmann, J., Jacobi, O., Mueller-Jentsch, W., 1975: Gewerkschaften in der Bundesrepublik, Studienreihe des Instituts fuer Sozialforschung, Frankfurt/Cologne: Europaeische Verlagsanstalt, 3rd Edition, Frankfurt/New York: Campus 1979.

Billerbeck, U., Deutschmann, Ch., Erd, R., Schmiede, R., Schudlich, E., 1982: Neuorientierung der Tarifpolitik?, Forschungsberichte des Instituts fuer Sozialforschung, Frankfurt/New York: Campus.

Brandini, P.M., 1975: Italy: Creating a New Industrial Relations System from the Bottom in: Barkin, S., (ed.). Worker Militancy and its Consequences, 1965-75, New York: Praeger, pp. 82-117.

Brandt, G., 1975: Gewerkschaftliche Interessenvertretung und sozialer Wanderl, Studienreihe des Instituts fuer Sozialforschung, Frankfurt/Cologne: Europaeische Verlagsanstalt.

Brandt, G., Jacobi, O., Mueller-Jentsch, W., 1982: Anpassung an die Krise: Gewerkschaften in den siebziger Jahren, Studienreihe des Instituts fuer Sozialforschung, Frankfurt/New York: Campus.

Braverman, H., 1974: Labour and Monopoly Capitalism. The Degradation of Work in the Twentieth Century, New York/London: Monthly Review Press.

Burawoy, M., 1978: Toward a Marxist Theory of the Labour Process: Braverman and Beyond, in:

Politics and Society, 8, pp. 247-312.
Burawoy, M., 1979: Manufacturing Consent. Changes in the Labour Process under Monopoly Capitalism, Chicago/London: University of Chicago Press.
Deutschmann, Ch., Schmiede, R., 1983: Lohnentwicklung in der Bundesrepublik 1960-1978. Wirtschaftliche und soziale Bestimmungsgruende, Forschungsberichte des Instituts fuer Sozialforschung, Frankfurt/New York: Campus.
Edwards, R., 1979: Contested Terrain. The Transformation of the Workplace in the Twentieth Century, London: Heinemann.
Friedman, A.L., 1977: Industry and Labour. Class Struggle at Work and Monopoly Capitalism, London: Macmillan.
Koehler, Ch., Sengenberger, W., 1983: Konjunktur und Personalanpassung. Betriebliche Beschaeftigungs-politik in der deutschen und amerikanischen Automobilindustrie, Forschungsberichte aus dem Institut fuer Sozialwissenschaftliche Forschung, Frankfurt/New York: Campus.
Koehler, Ch., 1981: Betrieblicher Arbeitsmarkt und Gewerkschaftspolitik. Forschungsberichte aus dem Institut fuer Sozialwissenschaftliche Forschung, Frankfurt/New York: Campus.
Lutz, B., 1982: Wirtschaftliche Entwicklung und Arbeitsmarktsegmentation. Entwuerfe zu einem in Arbeit befindlichen Buch, Munich, unpublished.
Mueller, G., Roedel, U., Sabel, Ch., Stille, F., Vogt, W., 1978: Oekonomische Krisentendenzen im gegenwaertigen Kapitalismus, Frankfurt/New York: Campus.
Nichols, T., Armstrong, P., 1976: Workers Divided, London: Fontana.
Nichols, T., Beynon, H., 1977: Living with Capitalism. Class Relations and the Modern Factory, London: Routledge and Kegan Paul.
Nobel, D.F., 1977: America by Design. Science, Technology and the Rise of Corporate Capitalism, Oxford: Oxford University Press.
Nobel, D.F., 1978: Social Choice in Machine Design: The Case of Automatically Controlled Machine to ls, and a Challenge for Labour in: Politics and Society, 8, pp. 313-47.
Parsons, T., 1966: Societies. Evolutionary and Comparative Perspectives, Englewood Cliffs: Prentice Hall.
Piore, M.J., 1983: Fissure and Discontinuity in U.S. Labour Management Relations, unpublished.
Regalia, I., Regini, M., Reyneri, E., 1978: Labour Conflicts and Industrial Relations in Italy, in:

Crouch, C., Pizzorno, A. (eds.), The Resurgence of Class Conflict in Western Europe since 1968, vol. 1, London: Macmillan, pp. 101-158.
Rubery, J., 1978: Structured Labour Markets, Worker Organisation and Low Pay, in: Cambridge Journal of Economics, 2, pp. 17-36.
Sabel, Ch., F., 1982: Work and Politics. The Division of Labour in Industry, Cambridge: Cambridge University Press.
Schmiede, R., 1983: Abstrakte Arbeit und Automation. Zum verhaeltnis von Industriesoziologie und Gesellschaftstheorie, in: Leviathan, 11, pp. 55-78.
Schmiede, R., Schudlich, E., 1976: Die Entwicklung der Leistungsentlohnung in Deutschland, Forschungsberichte des Instituts fuer Sozialforschung, Frankfurt: Campus.
Sohn-Rethel, A., 1978: Intellectual and Manual Labour. A Critique of Epistemology, London: Macmillan.
Streeck, W., 1979: Gewerkschaftsorganisation und industrielle Beziehungen, in: Matthes, J., (ed.), Sozialer Wandel in Westeuropa. Verhandlungen des 19. Deutschen Soziologentages Berlin 1979, Frankfurt/New York: Campus. pp. 206-226.
Zeitlin, J., 1980: The Emergence of Shop Steward Organisation in the British Car Industry: A Review Essay, in: History Workshop, No. 10, pp. 119-137.

Part Two

THE POLITICS OF RATIONALISATION: THE CAR INDUSTRY

Chapter Four

RATIONALISATION AND INDUSTRIAL RELATIONS: A CASE STUDY OF VOLKSWAGEN

Eva Brumlop and Ulrich Juergens

Some Basic Information on VW

Volkswagen has a special position in the industrial relations system in West Germany and among its automobile manufacturers. This paper describes how this special position affects the rationalisation policies pursued by Volkswagen in two key areas. Firstly, we consider how the highly professional, co-operatively orientated system of industrial relations which has emerged at enterprise level facilitates and limits rationalisation. And, secondly, we consider the durability likely of this system in the face of the growing problems which can be expected in the future.

VW is a multinational corporation. In 1980 it derived 26% of its total sales revenue from the activities of its foreign affiliates. Of the 2.5 million automobiles it produced in 1980, one million were produced abroad. Likewise, VW employed 159,000 workers in the Federal Republic and 99,000 abroad. In addition to car production, VW is also involved in manufacturing trucks, office-machines and computers, and robots. Nonetheless 75% of sales revenues in 1982 were attributable to vehicle production (including trucks) and VW still sees itself, and is thus seen by the public, as a West German company which produces automobiles.

Within the Federal Republic the VW corporation comprises two independent companies: Volkswagen-AG (hereafter the company or Volkswagen) with over 120,000 employees in 1982 and Audi (Audi-NSU-Auto-Union-AG) with more than 30,000 employees. The former company has six plants scattered throughout Northern Germany, three of which are devoted to assembly (Wolfsburg, Emden, and Hannover) and three to parts production (Kassel, Salzgitter, and

Braunschweig). But Wolfsburg over-shadows all other plants in terms of employment (58,000 workers in 1982) and its strategic significance for labour policy and industrial relations.

Special Features of the VW Industrial Relations System

VW has a special position among West German car firms because of (a) its special product and locational structure, (b) its special public character, and (c) its special industrial relations structure. All three features largely stem from VW's origins in the Nazi period as a central "project" of National Socialist economic and social policy. These features were reproduced after even the war.

(a) Under Hitler's sponsorship, Porsche emphasised from the start the idea of the car as a mass product and VW has remained orientated economically to the mass market. Thus the original design and product philosophy expressed in the "Beetle" was retained until the end of the 1960s and primary emphasis was given to the development and expansion of markets and quantitive growth rather than to style-orientated product differentiation (cf. Schnapp, 1979). In addition the Nazis selected a plant location in a structurally weak agricultural area for military-strategic reasons and this locational emphasis has also continued. This is reflected in the structural legacy of an integrated self-sufficient factory system modelled after the River Rouge complex in Detroit. Finally the original product conception and ideological orientation are reflected in the continuing public image of the VW as a qualitatively excellent and reasonably priced car for the average person.

(b) The special influence of the state also results from VW's roots in the Nazi period and survived through the British military administration to the Federal Republic. Not even the partial reprivatisation of the firm through the issue of "Volksaktien" (people's stocks) in the early 1960s changed VW's character as a state enterprise. The federal government and the Land government of Lower Saxony (each of which holds twenty per cent of the outstanding stock) retained their dominant influence in VW's supervisory board. This enabled the state representatives to form a coalition with those from the employee and trade union side and thereby to determine the composition and policies of the firm's management. Such a constellation existed during the early to middle 1970s, for example, when both the Land and federal representatives were sent by Social Democratic governments. During this period one continually heard cries for or against the "politicisation" of decisions at the so-called "Red Firm".

One should not exaggerate the state's dominant position but it is especially clear in two areas. Firstly, there have been attempts, particularly by the government of Lower Saxony, to utilise VW in the context of regional policy. Thus Volkswagen, in order to compensate for differences in employment levels across plants, has shown much flexibility in transferring employment between plants and has sought a proportionate development of jobs even where this has sometimes meant preferential treatment for individual plants and led to conflict between them. Secondly, the state has also influenced the form of industrial relations at Volkswagen. It should be emphasised that the special role of the state at VW need not always work to the advantage of the employees and trade union. Indeed, with the election of conservative regimes at federal and Land levels, there has been a clear change of climate at the firm. The works council and the trade union are increasingly on the defensive in wage issues and "generous" social regulations at VW have been criticised.

(c) The industrial relations situation at Volkswagen differs from other large industrial firms in two main respects: (i) the conduct of collective bargaining and (ii) how the institutions and bargaining systems of the centralised interest representation and interest mediation are attuned to each other. We deal first with collective bargaining.

In contrast to the usual system of regional collective bargaining, the VW company negotiates its own firm-level collective agreements. This makes it possible to conduct collective bargaining which is more specific to the automobile industry and VW's own situation. Indeed it is sometimes possible to agree on special regulations which the metal-working industry employers' association (Gesamtmetall) would otherwise refuse and which are sometimes controversial within the corresponding trade union (IG-Metall). In this respect the situation at VW is closer to a company-based collective bargaining system, such as that found in Japan and the USA.

The system of interest mediation and

representation is also distinctive. For the central works council and the trade union (IG-Metall) are both represented on the supervisory board by their respective chairmen and the union leader is traditionally its vice-chairman. High-ranking representatives of the Land and federal governments and representatives of banking and industrial interests also sit on the board alongside those from top management. This creates a tripartite communication structure which can be used to reconcile interests in dealing with the central policy issues of the firm.

The 1976 amendments of the Co-Determination Law, which provide for the establishment of the office of director of labour relations (Arbeitsdirektor) on the management board, created an additional level of centralised professional problem-solving for personnel issues at Volkswagen. As a member of the firm's top management the director for labour relations is directly involved in the operational decisions of top management; and, owing to the way he is chosen, he is orientated towards achieving compromise between the interests of capital and labour. In contrast to other West German automobile firms, the adoption of this institution from the co-determination model in the iron, steel and mining industries (Montanmitbestimmung) took place at VW without any problems. The current occupant of the office was previously employed by the labour union IG-Metall and edited its theoretical journal ("Gewerkschafter") and can use his good ties to the union for informal communication and mediation.

The most important institution in the system of bargaining levels described here is the Works Council. Because collective bargaining agreements are negotiated for Volkswagen at the level of the firm, the dualism of firm level and labour union interest representation which is usually considered to be an essential factor explaining the co-operative attitude and pattern of industrial relations hardly applies, if at all, to Volkswagen.

The leading members of the works council also participate in the collective bargaining negotiations. Thus Volkswagen also has a special position within IG-Metall. The plant and collective bargaining regulations and procedures at Volkswagen sometimes provoke controversy within IG-Metall especially with respect to rationalisation policy. The existence of overlapping personnel - the chairman of the works council of the firm is at the same time a high official of IG-Metall - guarantees that the differences remain limited and under control.

The works council plays a dominant role in the system of industrial relations at VW. Its influence even extends into the sphere of management. This is most evident in personnel selection at VW. For the works council has a considerable influence on the career patterns through its role in appointing foremen, superintendents, and white-collar workers. Moreover, because the works council is represented on the supervisory board, it also influences management selection. All this creates a middle level of interest mediation and overlapping interests below the level of top management at VW.

Area works council members and union work-place play a key role in mediating between the plant and company works councils and the shop floor. Indeed union stewards have a key role at shop floor level because there is ordinarily one union steward for every twenty-five employees compared with one works council member for every 700-800 employees at Wolfsburg. Together area works council members and union stewards constitute an efficient system of interest representation at shop floor level and they can interpret and apply centrally negotiated regulations and agreements at this level. Certainly there is every indication that the above-mentioned formal system and its results are accepted as legitimate reflections of workers' interests.

Thus 91% of the workers at Wolfsburg belong to IG-Metall - which is similar to the average in comparable car plants in West Germany. Among white-collar employees the union density is twice that of comparable car plants (81% in the case of Wolfsburg). In the most recent elections for the works council (1981 and 1984) IG-Metall won 85% of the seats. Oppositional candidates have made no inroads so far (on the composition of the general works council, see: Co-Determination in Volkswagen-AG, 1982).

There has never been an official strike at Volkswagen. Indeed, although the system of company level collective bargaining makes it possible to win collective bargaining demands through strikes without having to consider the weakness of small and medium-sized firms in the metal working industry outside the automobile sector, this has never been done. Likewise, there are scarcely any reports of informal strikes, walkouts etc. The typical form of collective protest at VW occurs when a group itself as being adversely affected by management measures, or fears that this might happen. It then makes use en masse of the right to take its grievance to the works council and thereby brings work to a halt.

To what extent, then, is the management authority at Volkswagen limited by the structure and functioning of the system of industrial relations? As the above description makes clear the system of industrial relations at Volkswagen has become similar to that in the iron, steel and mining industries. This "Montan" model (see also Ritz, 1983) is also based, on the one hand, on highly formalised procedural forms prescribed by law and, on the other hand, on a high degree of consensus between works council and management about the forms and content of co-determination. This consensus at the level of centralised forms of bargaining also provides a basis for informal levels of communication and procedure. In the context of this system, the works council controls central parameters of management decision-making.

Under these conditions autonomous forms of shop floor action such as restrictive practices (e.g. job demarcations) or the appeal to customs and practices have hardly developed. Empirical evidence about their frequency and importance is in fact quite limited (Dombois, 1982); such forms of behaviour have scarcely been a matter of concern for the unions and are frequently not considered in studies by industrial sociologists. It can be stated that at Volkswagen there is no tradition of autonomous shop floor action in the sense of a decentralised politics of the workplace in contrast to the union and plant level interest representation of the employees.

One possible explanation for the high level of consensus and successful co-operative regulation of conflict at VW is that the firm's economic prosperity has made it possible to "buy off" potential criticism and opposition by monetary concessions. This thesis seems quite plausible. An essential motive for working for VW, which is mentioned again and again in interviews with workers, is the high wage level; and it is true that, because of the geographic location of VW plants in either little-industrialised regions or in regions with declining industries and high rates of unemployment, VW pays wages at least 30% higher than other employers in its plant locations. Corresponding to this, the personnel cost structure is always referred to as the "Achilles heel" of VW. The strong position of the trade union in Wolfsburg, it is argued, keeps personnel costs so high that all efforts at rationalisation and progress in production techniques are eaten away.

But this line of argument is incorrect. The average effective earnings which a worker on incentive pay received in the Wolfsburg or Hannover plants during the second half of the 1970s was only average for a sample of eleven plants of West German automobile manufacturers (Hildebrandt, 1981). Only since 1980 are earnings at VW a little above average for hourly-paid workers (ca. 9%). With respect to fringe benefits, while Volkswagen is in many areas in the top rank among other automobile manufacturers, it is not particularly outstanding. Daimler-Benz is in many cases ahead of, or ranks equally with, Volkswagen.

The Introduction of Industrial Robots and the Response of the Works Council

Two phases in the technical and organisational development at VW may be distinguished. The first is characterised by orientation towards a single product and mass production in very large quantities, which lasted until the mid-1970s. The second phase involves the transition to a more differentiated product range in relatively small and medium numbers; and has come into full swing since the beginning of the 1980s.

The second phase of technical organisational rationalisation took place against a background of changed economic circumstances. Changed market requirements (quantitative and qualitative change in demand, the need for product diversification and a shortening of the cycle of innovation, etc.) could no longer be satisfied through the rigid automation techniques and relatively inflexible, "Taylorist" forms of work organisation, which had developed under conditions of highly standardised mass production. The company was increasingly compelled to develop and introduce manufacturing methods and technologies which could be flexibly applied to a variable product mix according to changing demand requirements and which was also as efficient and cost-saving as possible.

With the introduction of industrial robots and computer control systems as a way of achieving flexibility in production, VW is now seeking a solution in production and organisation technology that makes it possible to reconcile changed market conditions with production requirements. They are the central elements in the present restructuring process in the various areas of automobile production at VW and they have far-reaching consequences for the work and employment situation.

The first large-scale utilisation of industrial robots took place in 1979 in the body shop of the van plant in Hannover (see Benz-Overhage, et al., 1982). In the course of introducing a new model, industrial robots integrated in a computer-controlled assembly line took over the task of spot welding of body parts, which had previously been done manually on a conventional assembly line. A total of 63 robots were used; this was at the time the largest and most concentrated use of robots in a German automobile plant.

The second wave of robot installations took place in 1980 in the course of the introduction of new passenger car models (Jetta, Passat, Polo, Derby, Audi). Thus, by April 1983, there were around 942 robots being employed in the eight domestic plants of the Volkswagen Corporation (Peipe et al., 1982), and Volkswagen plans to install two thousand by the year 1990. They are now principally being installed in the area of tool handling (81%), mainly for spot welding and to a lesser extent arc welding and paint shop operations. Apart from the van line in Hannover, industrial robots have so far primarily replaced conventional production equipment (dedicated mechanisation). Their introduction has not, therefore, essentially increased the previous level of mechanisation, and job losses remain limited. Indeed the reduction in jobs as a result of the rationalisation measures has been partly offset by the expansion involved in growth of the model range and variety of optional extras and accessories offered (Peipe et al., 1982). The increasing volume of labour together with provisions in collective agreements providing for reductions in working hours (see below) have hitherto caused the demand for labour in assembly operations to increase steadily and made it possible to offset the jobs lost through rationalisation elsewhere through internal transfers.

In the course of introducing the new Golf model Volkswagen began to use robots in the area of labour-intensive final assembly operations in the summer of 1983; it is the first automobile producer in the world to use robots in large numbers in this area (70 assembly robots). Operations in final assembly have until now been most resistant to automation because of the multiplicity of tasks and the variability of the parts involved. Assembly robots are now engaged in the mounting of sub-assemblies as well as in the installation of hard trim (bumpers, wheels, etc.). Final assembly operations, in which today nearly 40% of VW's direct production personnel are engaged and which, together with paint shop operations, accounts for 50% of the total labour in manufacturing (Ehmer, 1983) will also remain relatively labour-intensive in the future. However, new technological developments, in particular in the area of tactile and optical sensors as well as flexible supply and stacking systems, combined with improved construction of the parts to be mounted, will increase the possibilities for rationalisation in assembly operations. The degree of automation in this area was under 10% so far in contrast to 75% to 80% in stamping and body shop operations; in future a considerable reduction in jobs as well as far-reaching changes in production techniques and work organisation can be expected. The introduction of the new automated car assembly operations at the Wolfsburg plant raised the level of automation to ca. 25% and has already made possible a reduction of ca. 1000 in manpower requirements. Moreover, 10,000 workers must be transferred in Wolfsburg alone in connection with the reorganisation for the production of the new Golf.

Mechanisation has also brought about a very important change in the structure of operations: manual assembly operations are now carried out largely by assembly robots. The elimination of jobs is only compensated to a limited extent by a small number of newly created jobs such as monitoring complex transfer line equipment (Anlagenfuehrer) and tool setting, besides which only simple residual operations (manual operator, feeder etc.) can be found. The latter predominantly involve "stop gap" operations within the partially automated manufacturing process.

According to statements made by the head of technology planning at Volkswagen, the rationalisation measures which the company has already carried out or plans to introduce include, in addition to the automation of manufacturing processes sketched here, the increasing use of computer technology in other areas such as research and development, design planning and production control. Within the next twenty years, "assuming a constant production schedule and allowing for a certain amount of product enhancement, a personnel reduction of 20-25% can be expected" (Ehmer, 1983). The structural implications of this process are a further reduction in machining, handling and assembly operations with, on the other hand, only a slight increase in machine control operations. The works council now strongly doubts whether this process can be offset by the increase in the volume of labour without redundancies.

The Response of the Works Council to the Rationalisation Policy and its Role in the Planning and Introduction of Industrial Robots and Data Processing Systems

In principle the works council has supported the rationalisation process pursued by the company since the middle of the 1970s as securing the future of the firm through better products and improved productivity. It hopes that jobs can be maintained by means of an improved position in world market competition (Peipe et al., 1982). Nevertheless, the works council is also concerned to prevent or offset the negative consequences of the rationalisation programme for the work, employment and income situation of affected employees. The demand for "social control" of the new technologies, which is formulated in a time of increasing mass unemployment, seems to express both an increasing scepticism towards the present technological change as well as an awareness of the need for an active and specific representation of employee interests in order to protect them from the negative consequences of rationalisation. Thus, the policy of the works council is characterised by supporting the modernisation policy at the same time as trying to make it "socially bearable" (Ehlers, 1983). This policy is supposed to encourage a compromise between economic profitability and "social progress" and the amelioration of the risks of rationalisation for the employees. The works council regards <u>co-determination</u> in the planning and use of new technology as well as <u>plant and collective bargaining</u> regulations on job design, working conditions, and protection of employment and income as the most important approaches to the realisation of its demands for "social control".

Negotiations between management and the works council take place in a basically "co-operative" manner, and in many areas, the works council enjoys rights of consultation and co-determination which go beyond what is legally required. It is thus in a position to take an active and early influence on the formulation of goals and decisions. Nevertheless, the real influence of the works council on the introduction of new technology seems to have been rather limited in the past. One explanation given by the works council is that its rights in the planning stage are merely limited to the right to receive information and be consulted. Since it lacks co-determination rights it lacks the required clear legal basis for effective action. While the works council is as a rule informed about planned changes in product and production technology, this information is not always so timely and comprehensive that suggestions from the works council should be considered. This criticism should be qualified by the observation that the works council did not always use the information and consultative rights at its disposal in the past to develop its own proposals for work organisation in a timely and effective manner.

These shortcomings of the "system of information and consultation" have led the VW works council to provide itself with more concrete and radical possibilities for influencing the introduction of new technologies. It now makes greater use of the advisory and negotiating committees, advisory staffs, and plant agreements. Thus in recent years project committees have been formed in the VW central works council as well as at plant level. These are concerned especially with issues of rationalisation and restructuring plans and include both a Planning Committee (Planungsausschuss) and a System Committee (Systemausschuss). The Plannning Committee so far exists only at plant level and gives information and advice to the works council about the planning of plant layout and materials flow and about the introduction of new facilities and machines. The Systems Committee, which was formed in 1980 and operates at company and plant levels, mainly discusses and negotiates problems associated with the introduction and control of data processing systems. According to the topics being considered the discussions of the works council committees also include representatives from management (e.g. personnel and social affairs, production planning, systems analysis).

Advisory staffs have also been established to take advantage of the legal rights to information and consultation. These provide the works councils with the expertise necessary for negotiations with management, evaluate company statements and develop alternative proposals and concepts to counter the negative social consequences of rationalisation plans. They are generally recruited from individual departments and areas of the firm and have access to inside information. This provides the works council with the necessary competence for conflicts with the firm, but it also entails a tendency towards "technocratic" types of solutions.

In addition the works council is making efforts to regulate the information procedures by supplementary plant agreements. They are supposed to ensure that the works council is informed, in accord with the provisions of the Works Constitution Act, in a timely, comprehensive and accurate manner, so that any possible suggestions for changes or other demands can be considered. In particular the information and advice should make clear which changes are intended in technology, organisation and/or personnel or how these are to be carried out in order to make possible a timely assessment of the expected consequences. The social and personal risks which are inevitably associated with the introduction of new technologies are not, however, thereby prevented. Faced with rapid technological change at VW, the central works council is calling for extensive amendments to the existing plant agreements in order to prescribe general conditions for the necessary regulation of social and personnel impacts.

This means that the approval of the works council must be given in negotiations with representatives of the company over the introduction of new technology. If the works council refuses to give its assent, either because the possible personnel effects have not been adequately presented or because no agreement can be reached on compensatory measures, then the installation of the system can be considerably delayed. For this reason, for example, the installation of a CAD system at VW was blocked for a year and a half.

Although the works council clearly attempts to fully utilise the existing legal framework by means of supplementary plant agreements and to win co-determination rights even in the planning stage of the introduction of new technologies, its real influence - in comparison to other fields of joint consultation - still seems to be quite limited in terms of gaining influence on the introduction of new machines and computer systems. This can primarily be shown in the case of the introduction of new production technologies such as robots. Here the effective intervention of the works council concerning questions of job design, qualification, working-conditions etc. has thus far been mainly limited to demands for certain revisions in the ergonomic design, the environment of the work-place, or manning levels. This is shown not only by investigations of the participation of works councils in earlier rationalisation programmes (Benz-Overhage et al., 1982), but also by the case of the new assembly line automation in which the activities of the works council seem to have been concentrated primarily on adjustment measures in the area of wages and personnel policy. There have been only a few cases in which the works council has succeeded in getting its ideas and demands on job design and training measures accepted in the course of the introduction of new production technologies.

Patterns for Coping with the Consequences of Rationalisation

Since the first rationalisation protection agreement of 1968, a policy of personnel adjustment has developed at VW in the course of the 1970s which is based on three principles:

(1) Personnel adjustments which are necessary as a result of rationalisation measures shall only take place under conditions which protect the acquired status (Bestandschutz) of the affected workers. In so far as this is not possible negative effects are to be stretched out over a longer time period; no dismissals are to take place as a result of the rationalisation.

(2) If a reduction in personnel is required by the economic situation, this should take place as far as possible through voluntary quits and the utilisation of natural attrition.

(3) The seasonal and short cyclical fluctuations in market conditions should take place without reductions in employment through adjustment of the volume of labour. A policy of hire and fire is rejected.

Closer analysis shows that these principles can only be maintained in crisis situations when certain social groups whose potential of resistance is lower than that of other groups can be pressured to leave "voluntarily". This applies mainly to internal plant "lines of segmentation" between German and foreign workers.

Since the beginning of the 1960s foreigners have been an essential labour market reserve for the German automobile industry. Though the percentage of foreign workers in the plants of the VW company, particularly in Wolfsburg, is traditionally less than in the plants of other West German automobile producers, personnel reductions in the crisis situation from 1973-75 were nevertheless disproportionately high among foreign workers. With a total reduction of 26% of the work force, the number of foreign employees was reduced by 68%. Of the total 33,000 employees who left Volkswagen from the beginning of 1974 until the end of 1975 over 13,000 were foreigners; the percentage of foreign workers sunk to a little over 7%. Since then this percentage has not increased significantly; and the remaining foreigners have meanwhile acquired considerable seniority in the plants. For this reason the potential for personnel reduction amongst foreigners is now considerably diminished.

In addition to the adjustment potential that can be achieved through the financial incentive of severence payments, usually three to four months' pay, independent of age, there is also extensive use of age-related early retirement. The latter instrument is particularly supported by the works council. The so-called "59 year-old regulation", which was agreed between the works council and management in 1975, stated that employees who had reached the age of 59 could accept voluntary redundancy. After one year of unemployment insurance payments, in which the firm itself pays the difference between unemployment benefits and the previous wage, the affected workers take early retirement under the social security system at the age of 60. In this way the firm was able to shift a considerable part of its personnel adjustment costs to public social funds.

The 59 year-old regulation was retained during the course of the 1970s even after the termination in personnel reductions. Among those "eligible" in each year about 80% have taken it. The advantages of this adjustment instrument are clearly so great that it was continued by the company even after the state stopped unemployment payment for those 59 year-olds who left employment in this fashion. Since then VW has paid the one-year of unemployment benefits out of its own funds. Under pressure from intensified rationalisation measures, the early retirement age was subsequently reduced to fifty-eight by a plant agreement. At the present time, discussions are taking place, at the urging of the works council, on the conclusion of a "57 year-old regulation".

The labour flexibility potential made available by these "voluntary" and "natural" forms of personnel reduction is quite large. Since the beginning of the 1970s the company has at no time had to engage in mass redundancies. Even personnel reductions amounting to more than a quarter of the entire labour force in 1974/75 were accomplished without redundancies, although management and the works councils had reached agreement on mass redundancies for 3,000 workers. The "voluntary" personnel reduction potential by means of severence payments proved to be greater than expected.

The 1974/75 phase represented a traumatic experience for all the participants at VW. A deep decline in sales coincided with the final end of the Beetle era and the transition to a comprehensively new production programme with uncertain market prospects. The fact that the restructuring of their production as well as the adjustment in personnel levels could be coped with so smoothly on a co-operative basis and, from the perspective of the subsequent developments, so successfully, led to a significant and long-term reinforcement of the pattern of industrial relations at VW.

Two general points are worth emphasising in considering the forms of regulation used in reducing the work force. Firstly, it is evident that personnel reduction, even on a large scale, can be processed in this system of industrial relations without a great deal of friction. And, secondly, company management and the works council reached a full consensus on carrying out personnel adjustments, individually "painless" and socially conflict-free. The works council participated in detail in determining and refining the measures through negotiations and agreements and generally accepted responsibility for them.

Strategies for Regulating the Volume of Labour

Naturally the works council would prefer to offset or reduce the pressure of redundancies rather than merely regulate how these redundancies are implemented. Since this pressure is not relieved at VW by blocking reorganisation or rationalisation measures as such, it is achieved through regulating the volume and intensity of labour.

The basic pattern in this area can be seen in the introduction of additional work breaks and the regulations for company recreation for workers on incentive pay (Leistungslohn), and educational leave for all employees. Break periods were first introduced in 1973 and have since been extended so that the actual number of daily working hours has been reduced to less than seven hours since 1981. According to the works council, this initiative protected 4,000-5,000 jobs at VW. But, contrary to the works council's expectations, it has not led to recruitment of a large number of additional workers owing to the increased pressure for redundancies involved in the rationalisation programme. The works council is none the less pleased that existing jobs have been saved in this way.

A further reduction in the volume of labour resulted from the regulations on recreation leave and educational leave. All employees in the German VW plants became entitled for the first time in May 1979 to a special leisure-time programme in addition to the regular six-week vacation, if certain prerequisites were fulfilled. A white collar employee who works a normal working day received, according to this plan, the right to ten recreation days leave every eight years; an employee on shiftwork receives ten days every three-and-a-half years; those who regularly work the third shift receive two weeks additional vacation every second year. There are similar regulations at other automobile producers but there is none in which it is so extensively used. Thus in the year 1981 20% of the work force at the Wolfsburg plant and almost 25% at the Hannover plant received an extra recreation leave in addition to their normal vacations. Comparable plants of other manufacturers give only 1-8% of their employees a corresponding recreation leave. The same is true of the educational leave which is sponsored by the Land of Lower Saxony since 1975 (courses and seminars on political and labour union topics and vocational training). Although this programme is open to every working person in Lower Saxony, it is only utilised extensively and systematically by Volkswagen. In 1982 16,000 employees of the Volkswagen company took the educational leave offered by Lower Saxony and in this way created 250 additional jobs in the Wolfsburg plant alone.

Strategies for Regulating Personnel Flexibility

The regulation of demands for greater flexibility in the use of labour constitutes a further issue in coping with the personnel consequences of the rationalisation process. The two most important agreements in this respect are the collective agreement concerning a new form of wage-differentiation (LODI) and the agreement on the maintenance of wages and salaries in the case of permanent transfers, especially as a consequence of rationalisation measures.

LODI replaces the system of analytical job evaluation, a highly complicated and differentiated procedure for wage determination which classifies jobs on the basis of a multitude of job requirement characteristics and their relative importance. In its ambivalence LODI represents a typical example of the way in which personnel problems resulting from rationalisation are handled. Its basic principle is the introduction of increased flexibility and the utilisation of labour in exchange mostly for enhanced income security (Hildebrandt, 1981).

The basis for evaluation is no longer individual jobs but certain work-subsystems ("Arbeitssysteme"), which as a rule include fifteen to thirty employees. Which jobs are included in a work subsystem is negotiated between management and the works council. The employees within a given work subsystem receive the same wage even in the case of slight differences in work activity and regardless of which job they are performing at a given time. As a rule, however, similar jobs, that is, at the same skill level, are joined together at a given work subsystem. This joining of previously separately evaluated operations not only corresponds to the interests of the employees in guaranteeing or raising their income (the wage level of the work subsystem is determined by the highest of the previous wage groups which is incorporated into the works subsystem) but also corresponds to the interests of management in more flexible deployment of labour. Temporary transfers that are necessary due to absenteeism, interuptions in production, etc., which in the past were always associated with friction, can now be accomplished much more easily.

Moreover, with the conclusion of this contract, further forms of flexibility were also agreed which provide for shifting workers between work subsystems within a manufacturing area or between areas. For a corresponding premium management receives relatively far-reaching powers of intervention and control over the assignment and use of personnel.

The second more traditional instrument for offsetting the consequences of rationalisation for individual employees or groups of employees is the regulations which have been signed in recent years between IG-Metall and Volkswagen for wage and salary maintenance in the case of permanent transfers. The limited guarantee of wages and salaries provides that any employee who is transferred to a lower paid job because of rationalisation measures, regardless of the length of time he or she has worked in the plant, will continue to be paid at the former wage level for at least 18 and up to 36 months. Furthermore, the plant is obligated to transfer him or her back to a job at the former wage rate in so far as that is possible. The ambivalence of this regulation consists in the fact that, on the one hand, personal income is guaranteed for a limited period independently of technical and organisational changes, on the other hand, down-gradings are legitimated, because it is shaped in a "socially acceptable" manner, experienced only individually and ameliorated by payment or compensations rather than being registered collectively. Regulation on wage and salary compensation in the case of work impairment protects employees from loss of earnings who, as a result of health limitations or increasing age, are no longer able to meet the performance requirements of their previous job; it also makes it possible for them to be transferred to a suitable job. This regulation is applicable to plant employees with at least ten years seniority, who are at least 50 years old or have had a work accident and whose entitlement to a transfer is confirmed by a recommendation of the plant physician.

The guarantee for older workers applies to employees who have worked at least ten years at VW and have reached their 56th birthday. They receive a guarantee of their former wage level without time limit in the case of transfers either for plant-related or individual-related reasons. An employee of at least 25 years seniority can, moreover, only be dismissed with the approval of the works council. It is characteristic for both these regulations that, on the one hand, they are instruments of protection against the consequences of rationalisation, at the same time, however, they give the company the possibility of raising the general performance level through the removal of older employees and those with diminished capacities. The fact that the regulation requires a minimum period of employment considerably weakens its effectiveness as a protective mechanism and encourages the practice of dismissing particularly those employees with low seniority.

In summary, with respect to this point it can be said that the works council has in principle accepted the interests of the company in increasing flexibility in the use of personnel and has been able to achieve far-reaching status protection with respect to wages. It has rid itself of a conflict potential which traditionally existed at shop floor level with respect to issues of temporary or permanent transfers and which previously involved union stewards and area works council members in day to day petty conflicts in which they were frequently subjected to criticism by the affected employees. Temporary and permanent transfers now take place in a rather routinised and legalistic manner between plant level supervisors and the affected workers.

Flexibility of labour deployment for workers with different job classifications, particularly at the borderline between skilled and unskilled work, also requires consideration. There are at present two tendencies at VW: the "enrichment" of production jobs by the addition of job elements that used to be performed by skilled workers and an increase in the percentage of skilled workers who are employed in production jobs.

As a result of job enrichment new job classifications are currently being developed such as monitoring complex transfer-line equipment in the body shop and final assembly. The tasks of the equipment monitor (Anlagenfuehrer) include changing electrode tips, dressing electrodes, minor robot programme adjustments etc. Another example is the setter-fitter (Einrichtungsmechaniker) in the stamping shop, whose job includes both the tasks of the die setter and mechanical repairman. The introduction of additional integrated job classifications is being discussed. The total number of workers involved is still small - there are at present about 350 equipment monitors at the Wolfsburg plant. Because pay is equal to that received by skilled workers, these jobs are eagerly sought. Although they are production jobs there is now a tendency to limit recruitment to skilled workers.

The enrichment of production jobs is not limited to the addition of tasks formerly performed by skilled workers; as in other firms of the automobile industry, quality control functions are being "returned" to production workers.

It has long been the practice at VW and other German automobile firms to employ qualified skilled workers in production jobs. Thus in 1982 40% of the wage earners at the Wolfsburg plant had completed an apprenticeship; the percentage is even higher in other plants (e.g. 56% at the Emden plant). This means that a large percentage of skilled workers have production jobs. Since most have qualifications as metal workers, i.e. relevant to the auto industry, there is a large reserve of skilled workers in the production departments that can be utilised by management in job enrichment strategies to achieve more flexible labour deployment. This percentage will increase in the future. As a result of the hiring freeze in effect since 1980, which will certainly remain in force, the only new recruits are graduates of the in-plant training programmes (about 1,500 per year). The recently introduced policy is to assign all these workers to production jobs after completion of their training without regard to their specific apprenticeship training; they can then - like all other production workers - apply for transfer to more attractive jobs, for example, in a skilled department. If this pattern of recruitment is maintained, the employees in VW auto plants will be predominantly skilled workers by 1990.

On the whole the co-operative response to intensified rationalisation at VW since the beginning of the 1980s is reminiscent of the model of the Japanese automobile plants. There is in practice a "lifelong employment guarantee", and no one is dismissed for other than personal reasons; the retirement age is dropping to about 58; the level of qualification of production workers is increasing, which makes possible various forms of increased flexibility in labour deployment. Greater employment security is closely tied to increased flexibility in labour deployment.

We have tried in this essay to describe concretely the functioning of the system of industrial relations at VW in order to present a differentiated picture of what crisis management on a co-operative or participatory basis between management and labour (Sozialpartnerschaft) in the Federal Republic really means. It is a system which has been able to reproduce itself with remarkable stability even under conditions of increased stress during the 1970s.

REFERENCES

Benz-Overhage, K., E. Brumlop, Th.v. Freyberg, Z. Papadimitriou, 1982: Neue Technologien und alternative Arbeitsgestaltung, Frankfurt.

Dohse, K., U. Juergens, H. Russig, eds., 1983: Aeltere Arbeitnehmer zwischen Unternehmensinteressen und Sozialpolitik, Frankfurt/New York.

Dombois, R., 1982: Die betriebliche Normenstruktur. Fallanalysen zur arbeitsrechtlichen und sozialwissenschaftlichen Bedeutung informeller Normen im Industriebetrieb, in: K. Dohse, U. Juergens, H. Russig, eds., Statussicherung im Industriebetrieb. Alternative Regelungen im internationalen Vergleich, Frankfurt/New York.

Dombois, R., 1982: Arbeitsplatz Volkswagen, in: R. Dombois, R. Doleschal, eds.: Wohin laeuft VW? Die Automobilproduktion in der Wirtschaftskrise, Reinbeck b. Hamburg.

Ehlers, S., 1983: Rede gehalten auf der Betriebsraetekonferenz der Volkswagenwerk AG in Kassel, Wolfsburg.

Ehmer, H.J., 1983: Neue Produktionstechnologien im Automobilbau, Vortrag gehalten auf der Betriebsraetekonferenz der Volkswagenwerk AG in Kassel, Wolfsburg.

General Works Council of the Volkswagen AG, ed., 1982: Co-Determination in the Volkswagen-AG.

Hildebrandt, E., 1981: Der VW-Tarifvertrag zur Lohndifferenzierung, Veroeffentlichung des Internationalen Instituts fuer vergleichende Gesellschaftsforschung, Berlin.

Mickler, O., P. Kalmbach, R. Kasiske, F. Manske, W. Pelull, W. Wobbe, 1981: Industrieroboter. Bedingungen und soziale Folgen des Einsatzes neuer Technologien in der Automobilproduktion, Frankfurt/New York.

Peipe, H., M. Pusch, B. Schroder, H.-J., Uhl, H. Ziegler, 1982: Robotereinsatz in der Volkswagenwerk AG aus der Sicht der Arbeitnehmer, Wolfsburg.

Ritz, H.-G., 1983: Betriebliche und staatliche Arbeitspolitik - am Beispiel der beruflichen Integration Schwerbehinderter, in: U. Juergens, F. Naschold, F., eds., Arbeitspolitik - Materialien zum Zusammenhang von politischer Macht, Kontrolle und betrieblicher Organisation der Arbeit, Special Issue of Leviathan.

Schnapp, J.B., et al., 1979: Corporate Strategies of the Automobile Manufacturers, Lexington/Toronto.
Streeck, W., A. Hoff, 1982: Industrial Relations in the German Automobile Industry: Developments in the 1970s, Discussion Paper IIM/LMP 82 - 25, Science Centre, Berlin.
Thimm, A.L., 1976: "Decision-Making at Volkswagen 1972-1975", in: Columbia Journal of World Business, pp. 94-103.

Chapter Five

THE POLITICS OF TECHNOLOGICAL CHANGE AT BRITISH LEYLAND

Harry Scarbrough

Introduction

This paper focuses on the introduction of new process technologies at British Leyland's Longbridge plant before the launch of the Mini Metro model in October 1980. Apart from the relevant literature and documentation, it draws upon a number of interviews with managers and trade unionists carried out at the plant between 1978 and 1981 (Scarbrough, 1982).

This study suggests that even at the most detailed level the process of technological change is inextricably intertwined with the politics of workplace industrial relations, and in particular with the pursuit of efficiency targets, the determination of work practices, and the distribution of skills in the production process. Moreover, the political conflict surrounding technological change is seen to occur upon several different planes; not at the workplace alone, but also, for instance, on the level of corporate strategy where important decisions are taken regarding the investment in new technologies and the distribution of costs and benefits arising from their introduction.

A Brief History of British Leyland

Although the British Leyland Corporation only dates back as far as 1968 with the government-sponsored merger of British Motor Holdings with Leyland Motors to form British Leyland Motor Corporation (BLMC), its origins can be traced back to the very beginning of the British motor industry. As the last British-owned volume car producer, BLMC was the end product of a long-term process of concentration in the

British car industry, the legacy of what were once over 60 independent companies. Beginning with the pre-war growth of Austin and Morris, and continuing after the war with their amalgamation in the British Motor Corporation, this process was carried forward in the 1950s and 1960s by a series of mergers and acquisitions which consolidated Jaguar, Rover, Standard-Triumph and finally Leyland Motors into the massive but rather ramshackle BLMC empire. The Longbridge operation itself was the biggest plant in BLMC - indeed the biggest in the British motor industry - having been founded in 1906 by Herbert Austin.

On its creation, BLMC with a total of no less than 60 manufacturing plants was the second largest car producer outside the U.S.A., and enjoyed a commanding 40% share of the U.K. market. However, far from achieving the expected economies of scale and greater efficiency, the new organisation quickly proved itself to be incapable of producing a coherent model range, let alone of solving the more fundamental problems of the British sector of the industry.

By 1974 the combination of internal weakness and adverse external pressures brought the firm to the verge of bankruptcy, and prompted a government rescue operation. The newly-elected Labour government viewed the reconstruction of the firm as strategically vital to the British economy - not least to the components industry which was heavily dependent on BLMC - and opted to take it into public ownership under the aegis of the National Enterprise Board. A team of Inquiry was appointed under Sir Don Ryder to formulate a long-term strategy for the Corporation.

The Ryder Plan and the Metro Project

The Ryder report published in April 1975 outlined a ten year strategy for British Leyland (BL). Based to a large extent on BLMC's own projections, the main thrust of this strategy was expansionist, forecasting an increase in production to 1,250,000 cars p.a. and a U.K. market-share of 33%. Reconstruction of the company was to be implemented through a comprehensive programme of modernisation and rationalisation. Years of under-investment were to be rectified by a £2,000 million investment programme, of which up to £900 million was to come from the government - the remainder being generated by BL itself.

A seemingly more intractable problem, however, was BL's chronic industrial relations malaise. A loss of management control due to tight labour markets and piecework payment methods in the 1950s and 1960s had only been exacerbated by the introduction of Measured Day Work (MDW) payments in the early 1970s. Shop stewards in BL were more effectively organised and more powerful in their own right than any similar group in the rest of British industry, and despite management's hopes MDW had done little to erode their power.

Ryder's prescription for BL's industrial relations problems took due account of the prevailing distribution of power by recommending the development of industrial democracy within the company: "The most crucial factor in improving industrial relations at British Leyland and in creating the conditions in which productivity can be increased is, however, that there should be some significant progress towards industrial democracy" (Ryder, 1975, p. 37).

The emphasis here was upon achieving the co-operation and commitment of stewards and workers alike in improving BL's economic performance - in particular, by reducing the number of unofficial disputes in BL plants and thereby improving continuity of production.

Given its aims and origins it was not surprising that in instances such as worker participation the Ryder Report reflected government philosophy. But of longer term significance was the Report's tendency to lean towards a particular perception of the "national interest" in developing a workable strategy for BL. This occurred despite the fact that the government had no national policy for the U.K. industry as a whole, and had restricted the extension of public ownership to the weakest sector of that industry - leaving the three American-owned multinationals to their own devices. Only at BL was corporate strategy allied to political and social considerations. The planned growth in production, for example, although it served to alleviate the social and economic impact of rationalisation, also managed to side step some difficult and perhaps necessary decisions on manning levels and plant closures. In that sense, one of the consequences of government intervention was to instil an element of political rigidity into BL's structure and operations: a rigidity which was implicit in dependence on government funding and the link with the national interest, but which did prevent the firm's piecemeal breakdown and absorption into one of the multinational producers (as was later to happen with Chrysler U.K.).

The fusion of BL's interests with the national economic interest is apparent even in one of Ryder's more specific recommendations - namely, approval for the development of a small car replacement for the Mini. This model - code named ADO 88 - was subsequently to take shape as the Mini Metro. Although the commercial rationale for a new small car was debatable - indeed was later to be challenged by at least two chief executives of the corporation - the Ryder team paid less attention to the low profit margins associated with small cars than with the implications for Britain's balance of payments:

> If BL were to opt out of this sector of the market, there would undoubtedly be a very substantial increase in car imports ... It ought therefore to be both in BL's interest and in the national interest to remain in the small/light sectors of the car market (Ryder, 1975, p. 16).

Despite its questionable commercial value, the technological lead-time for projects of this sort meant that the ADO 88/Metro decision was one of the few Ryder recommendations to survive BL's subsequent crises relatively unscathed.

From 1977 to 1981, the Longbridge plant underwent an extensive £200 million programme of modernisation to prepare it for production of the Metro. A great part of this investment went into the construction of a new "Body in White" (BIW) shop, the "New West". The BIW process involved the spot-welding assembly of up to 200 sheet-metal body panels into a rigid unpainted bodyshell. The overall design of this process was shaped by Ryder's strategic objectives for the Metro. It exploited to the full the technological possibilities of mechanisation and automatic control. The lines of industrial robots and automated systems in the New West presented a sharp contrast with the labour-intensive and rudimentary "Box Jig" BIW methods used in the old West Works - a shop dating back to the pre-1st World War period. As such the BIW process represented the greatest technological advance of any of the new Metro facilities.

Technology and Job Controls

Although primarily orientated to strategic goals of volume and cost, it would be naive to suggest that the design of the BIW shop did not deliberately exploit the opportunities for increasing management's control of production. For the most part, however, management control was built into the BIW shop simply through the mechanisation of the production process. In the case of the Body-Framing sub-process, for instance, the area where sub-assemblies were spot-welded together to form a rigid bodyshell, management control was actually programmed into the two lines of 14 industrial robots performing this task, and the direct labour element completely eliminated. Body-Framing on the Mini required a total of 63 direct workers and 9 maintenance workers, but in the New West only 15 maintenance workers and one machine operator were needed.

A parallel can be drawn with the installation of 39 industrial robots in the BIW area of Ford's Halewood plant. Here the control rationale was explained by the Manufacturing Manager thus: "We haven't got control of the labour-force. We can't force each man to put each weld in the right place. So we've tried to build in quality through machines" (Interview, September 1980).

As this example illustrates, automation tends to transform social control into an abstract, built-in feature of the production system; where, for example, the introduction of industrial robots might be justified in the following terms: "a réduction in the number of random factors that influence the quality and efficiency of the process" (Owen, 1980).

Despite the extension of management control through mechanisation, even the new technologies of the BIW shop could not entirely eliminate the subjective element from production. Labour remained a pervasive, though increasingly marginal factor, with groups of direct workers installed at various interfaces in the BIW process to provide the still needed human attributes of skill and flexibility. Apart from the 193 maintenance workers (over two shifts) required by the automated facilities, large numbers of sheet metal workers were deployed on the finishing lines to perform smoothing and CO_2 welding operations.

The continuing presence of a manual labour-force gave the BIW innovation a qualitatively political dimension. The economic viability of the Metro BIW process depended on its attaining targets for output and labour productivity comparable with the new model's major rivals in the small-car segment. Yet, although management's efficiency targets had been built into the BIW process as design assumptions on line rates, conveyor speeds and machine efficiencies, the formal parameters of the BIW design could only evoke but not secure the appropriate, efficient or "necessary" work practices on the part of the manual labour-force. Even, or perhaps especially, in such a highly automated and integrated system, the manner in which the human work-force was incorporated into the production process became so critical as to demand both careful consideration and a coherent general policy from management.

It was clear, for instance, that if Longbridge management were to achieve the man-hours/car target planned for the New West, the establishment of comprehensive and unchallenged management control of shop-floor labour was a vital imperative. Its achievement, however, involved confronting and overturning the intricate network of custom and practice or political "constraints on efficiency" which had been established by a powerful union organisation at the plant over the previous twenty to thirty years.

In Unit One of Longbridge, which included the new BIW shop, a well-developed set of job controls were watchfully maintained by an active and troublesome Unit Committee of shop stewards. Significant inroads had been made upon management's control of production, as the following features indicate:

Labour Mobility	- mobility within a Production Unit only, and with the agreement of shop steward.
Industrial Engineering (Time and Motion, Work Study)	- industrial engineers' access restricted, having to be accompanied by shop steward. No access to maintenance tasks.
Industrial Engineering (Contd)	- A "mutuality" clause enshrined in Longbridge Direct Workers' Agreement (1975)

	meant that new or revised work standards had to be agreed with shop steward, with status quo applying until any dispute resolved.
Demarcations	- Unit One contained the only formally recognised demarcations on the Longbridge site; sheet metal finishing jobs being allocated to members of the Sheet Metal Workers' Union, and door-hanging and bootlid fitting to members of the TGWU.

Clearly, job controls of this sort - especially the mutuality clause - inhibited management's ability to deploy labour and determine manning levels, and thus represented a major threat to the economic and technical rationale of the new BIW shop. To remove this threat, the period 1977 to 1980 at Longbridge saw not only an extensive programme of technological change, but also a series of management initiatives aimed at a radical transformation in management-worker relations, and the establishment of "efficient" practices.

The Management of Change

The first stage in managment's pursuit of change was closely linked to the system of "employee participation" which had been developed within the company; significantly, in implementing the Ryder Report proposals BL top management had chosen this term in preference to Ryder's "industrial democracy". An elaborate structure of "joint management committees" had been established in 1976 at three major levels in the organisation - Plants, Divisions, and at the apex the Leyland Cars Council - with the terms of reference for the committees defined as follows: "... the main task of each body in each level of the Employee Participation System is to improve the performance of the activity within which the employees who are represented in the body are employed" (British Leyland, 1975, p. 4).

At each level a wide range of issues relating to company performance could be discussed by shop stewards, staff representatives and managers - at Cars Council level, for instance, product plans, facilities plans and capital allocation came into discussion, while plant level committees could raise questions on budgets, plant performance and production schedules. Despite the emphasis on improving production efficiency, the range and depth of employee involvement indicated that participation was more than just a manipulative exercise by top management. The participation structure was approved by mass meetings of the work-force at most BL plants - Jaguar at Coventry being a notable exception - and many stewards embraced it as a significant step forward in industrial relations. At Longbridge in particular, the senior stewards - many of them members of the Communist Party - saw the advent of participation at a corporation now under public ownership as an event of major political significance. Derek Robinson, the plant convenor and leader of the BL shop stewards combine, emphasised the wider implications of the new system: "If we make Leyland successful, it will be a political victory. It will prove that ordinary working people have got the intelligence and determination to run industry" (Guardian, 9.4.1979, p. 19).

The neat, almost bureaucratic framework of participation contrasted greatly with the highly fragmented nature of collective bargaining within BL - by 1975 there were 246 separate bargaining units and 17 recognised unions in the company. Although participation tended to place certain limitations on the senior stewards - notably over their privileged access to information - the specific exclusion of negotiable issues from the scope of the new system meant that it exerted relatively little influence, either direct or indirect, upon the processes of collective bargaining. As a result, the participation machinery quickly became seen by many managers and trade unionists alike as irrelevant to the most crucial questions in management-worker relations.

While participation at company level was falling into disrepute, however, at the Longbridge plant it was able to exert a perceptible influence upon the social and technological progress of the ADO 88/Metro project. From November 1976 to September 1979, the joint ADO 88/Metro participation sub-committee provided a forum for Longbridge managers and shop stewards to discuss the changes in production and working method specifically related to the new Metro facilities. With its clear terms of reference and dynamic subject matter, the committee's discussions were viewed favourably by the Longbridge senior stewards. As Derek Robinson explained, the committee allowed the unions to "look objectively at some of the changes that were required outside of being in a bargaining position" (Robinson, 1980). Jack Adams, Robinson's successor as plant convenor, felt that it was "the most successful participation exercise that we took part in ... less abstract than other exercises" (Interview, 28.8.1980).

Despite favourable comments of this sort, the influence which the stewards on the committee exerted on the technological changes for the Metro seems to have been comparatively marginal. Apart from lacking the kind of technical expertise needed to properly evaluate the new systems, the stewards' contributions in committee understandably tended to centre on issues such as job content and working conditions - questions essentially peripheral to such large-scale complex technologies. Nor did the Longbridge stewards challenge the economic priorities which were at the heart of the design process - seeing features such as "job enrichment" as necessarily having to "fit in" with a production system defined by considerations of efficiency.

The trade unionists' acquiescence in the economic realities of the design process - in itself hardly surprising - was compounded by a willingness on the part of some of the management team to define the whole ADO 88/Metro participation exercise in purely instrumental terms. Comments such as "participation was a good P.R. exercise" or "participation was a success and helped us in selling the changes we wanted" from some of the managers on the committee may be seen, however, as simply reflecting an honest appraisal of the basic rationale behind participation at BL - worthwhile to the extent that it helped to achieve managerial objectives. Certainly, "salesmanship" and "P.R." seemed to come to the fore in the managerial effort to link the changes in technology at Longbridge to a broader programme of changes in work practice, including the dismantling of demarcations and ending of "mutuality".

Management's first such attempt came in the post-Ryder period when, nourished by the relative openness of participation, a consensus had developed at Longbridge on the need for major changes in work rules and practices. With the Works Committee of Longbridge stewards ready to countenance radical changes in custom and practice on the condition of "payment for change", plant management responded with an innovatory set of proposals - including elements of job enrichment in a so-called "team working" concept - which promised to secure a degree of labour flexibility comparable with Japanese practices. The proposed production "teams" of 20-30 workers would "accept responsibility for quality of their own work", and each would include multi-competent "adjusters"; workers who would perform rectification jobs on the track, having gained a range of "skills" through some form of "job rotation". It was management's expectation that over time the majority of team workers would acquire this degree of flexibility. Discussions on these proposals, conducted on what was virtually a productivity bargaining basis, lasted from 1977 to the summer of 1979, with the Works Committee issuing bulletins to its members to "condition" them for the changes to come.

In fact, however, Longbridge management's hopes of securing change by an agreement at plant level were never to be fulfilled - not because of the recalcitrance of the trade unions at the plant, but because of a change in management strategy at corporate level. When Michael Edwardes was appointed chief executive of BL at the end of 1977, it signalled a shift in strategy which was eventually to lead to the abandonment of Ryder's expansionist policy, and with it the goal of securing trade union co-operation and consent in the management of the company. A rapid decline in BL's U.K. market share (from 30.9% in 1975 to 23.5% in 1978), continued industrial relations problems, and, from April 1979, the presence of a hostile Tory government, all reinforced Edwardes' determination to implement a ruthless programme of rationalisation at BL, aimed at stemming the firm's increasingly disastrous financial losses. The corollary of this approach was an industrial relations policy which sought to regain what Edwardes saw as "the sine qua non of survival ... the right to manage."

It was axiomatic to this policy that it could tolerate neither the financial nor the political implications of the kind of agreement being sought at Longbridge - in Edwardes' terms a "buy out" of restrictive practices - if only for the precedent that might be established at other BL plants. Thus in the summer of 1979 the negotiations at Longbridge were finally terminated, victim of a policy that one Longbridge manager described as "getting the rules of the game right across the company" (Interview, 30.1.1980). As this comment implies, BL's transition from a system of dependency and subsidy to one of independence under market conditions placed great emphasis upon the role and management style of the corporate leadership, Sir Michael Edwardes in particular. The urgency and severity of the transition demanded the centralised application of power - tending to counteract Edwardes' earlier preference for a more de-centralised organisation structure.

Having discarded participation and negotiation as instruments for changing shop-floor practices, the alternative adopted by Edwardes and his team was heavily conditioned by the financial and political crisis which beset BL in 1979 and 1980. Not only was the corporation - contrary to Ryder's plan - becoming more and not less dependent on government aid (as outlined in Table 1), but its market share was being steadily undermined by adverse exchange rates and an out-dated model range.

Table 1: State Funding for BL, 1975-84

		£ million
1975-76	Purchase of shares	46
	New equity	200
1976-77	Loan later converted to equity	100
1977-78	Loan converted to equity	50
1978-79	Equity	450
1979-80	Equity	150
1980-81	Equity	300
1981-82	Equity	620
1982-83	Equity	270
1983-84	Equity	100
	Further funding promised	100
	Wholesale Vehicle Finance for Dealers' stocks	25
		£2,411 million

Moreover, although the kind of rationalisation pursued by Edwardes promised greater financial stringency in the short term, it could not of itself secure BL's economic viability in the future. Major and permanent reductions in production capacity inevitably meant the loss of any useful economies of scale, and also effectively precluded the strategically efficient multinational configurations adopted by the major producers. Indeed, while BL was closing its last continental car assembly plant and withdrawing to a parochial U.K. base, multinational rivals such as Ford and General Motors were further undermining the corporation's home market by bringing in "tied imports" from their integrated European marketing and manufacturing networks.

Edwardes' response to such problems, however, put the highest priority on stabilising BL's financial position, and ensuring future funding by persuading a sceptical government of the Company's "viability". Since these aims could no longer be pursued through scale or internationalisation, BL management now placed great, almost desperate significance on achieving a substantial and rapid increase in labour productivity levels within the firm. Evidence for the imperative and urgent concern attached to this goal comes from BL's 1980 Corporate Plan. As Table 2 indicates, management's plans for the Austin-Morris division alone envisaged an increase in productivity of around 30% between 1980 and 1982, and within a very few years this improvement was planned to reach 150%. At Longbridge itself the goal was no less than a 100% increase in productivity after the launch of the Metro.

Table 2: BL Corporate Plan 1980: Austin-Morris, Planned Volumes and Manning Levels

	1979 (actual)	1980	1981	1982
Manning	42.6	39.6	34.5	34.4
Volume (000's)	461	407	449	464

The pursuit of productivity gains of this order and urgency, coupled with the restoration of management's "right to manage", boded ill for the kind of workers' job controls extant at Longbridge. By November 1979 BL management were engaged in an all-out attack not only against such "restrictive practices", but also against the work-place union organisation which sustained them. In carrying out this attack, however, BL management possessed few of the weapons which multinational rivals such as Ford could deploy - most notably the ability to threaten the transfer of production or investment. Yet even without this guaranteed margin of political control, it was a particular discovery of BL top management that beyond a certain point the company's market decline perversely tended to strengthen and not weaken their bargaining position. Where worker participation had clearly failed to impress market disciplines upon the BL trade unionists by institutional means, the very real threat of collapse posed by financial crisis and government policy served to impose financial stringency directly upon the collective bargaining process, while in equal measure boosting managerial authority.

In a context of steeply rising unemployment, the "rationalisation" of BL and the closure of all but two volume car assembly plants (Longbridge and Cowley) proved, as Table 3 indicates, an inherently potent force in cowing the remaining workers.

Table 3: BL, Manning and Internal Disputes, 1977-1981

	1977	1978	1979	1980	1981 (Nov)
Total U.K. employees (000's)	176	164	146	118	96
Reduction in year (000's)	-	12	18	28	22
Man-hours lost through internal disputes (millions)	14.8	10.0	4.8*	3.5	0.9

* Excluding man-hours lost in CSEU strike affecting all engineering workers

Moreover, by exploiting BL's dependence upon a government anxious to maintain a strictly arms length relationship, Edwardes and his Board were able to achieve something of the multinationals' ultimate veto by, in effect, installing themselves as primary arbiters of the company's future. On those occasions when the eleven BL unions sought to organise serious opposition to management's plans, Edwardes was able to coerce their members back to work by the simple expedient of publicly threatening the mass dismissal of strikers or even the liquidation of the whole company - tactics which no private sector chief executive could have employed, if only because of the likely effect on share price.

Neither threats of mass dismissal nor the withdrawal of government aid were entirely novel at BL - the first having been used in 1970 over the introduction of MDW at Cowley, and the second by the Labour government in the mid-1970s in response to continuing industrial relations problems. However, it was only in the particular circumstances of the late 1970s, and especially the government's adoption of a policy of disengagement, that such threats became perceptibly real - allowing management to exploit the power inequality inherent in the employment contract as a weapon against collectivities of workers. This kind of coercion combined with judicious use of secret ballots of the work-force brought about a transformation in the climate of industrial relations within BL. Although this did not amount to the "change in attitudes" sought by Ryder, the effect was such that management were even able to secure the dismissal of Derek Robinson on the pretext of his opposition to the Edwardes policy. When the Longbridge work-force refused to support Robinson's reinstatement, their vote symbolised not only the crushing of union power in BL but also the alienation of rank-and-file sentiment from their shop steward leadership. More importantly, it paved the way for the company-wide imposition - again under threat of dismissal - of the so-called "Draft Agreement" in April 1980.

The Draft Agreement was a 92 page document which swept away all existing agreements and custom and practice in BL plants, and replaced them with a comprehensive set of work rules specifically orientated towards the operation of new technologies, and designed to guarantee an almost unfettered control of the shop-floor labour-force. It included the following clauses:

Productivity : "the introduction of sound working practices and elimination of ... restrictive practices, and of all other constraints upon effective operating."

	:	labour to be mobile throughout plant and to perform any grade of work within ability.
Work time	:	"the unions undertake to ensure that no restrictive practices will be applied to overtime, e.g. 'one in, all in', guarantee of overtime etc."

<u>Longbridge and the Edwardes Regime</u>

With the impending launch of the Metro and high productivity targets to achieve on the new facilities, Longbridge management were especially vigorous in applying the Draft Agreement. Not only did it contain almost all the items previously sought as a negotiated settlement, but it served as the "Enabling Bill" for production managers throughout the plant to implement their own "shopping lists" of detailed changes in work practice. Mutuality was finally ended, and the determination of manning levels and work standards became the prerogative of BL's industrial engineers. With the ending of tea-breaks and installation of a corporate framework for relaxation allowances, the two-shift working week was increased from 72.6 to 78 hours.

Meanwhile, the plant's union organisation was subjected to an even more overt attack, as the Unit committees of stewards were disbanded and the Works Committee members deprived of their semi-permanent status, leaving only the convenor and his deputy as full-time worker representatives.

In the face of these attacks, the trade unionists at Longbridge remained strangely passive. Successive defeats over Derek Robinson and the Draft Agreement notwithstanding, the programme of facility changes at the plant offered the unions numerous opportunities to at least hinder management's plans whilst avoiding full-scale confrontation. That the Works Committee did not exploit such opportunities seems to be attributable, in the first instance, to the implications of their policy towards new technology. Having, through the BL Combine, made "massive investment" the key element in their alternative policy (a set of demands strongly influenced by Communist Party thinking), the Longbridge shop stewards were in a sense deeply compromised when that investment actually materialised at the plant. It became impossible for them to "irresponsibly" obstruct the process of technological change and the commissioning of new facilities.

Second, and more importantly, the Longbridge stewards were acutely aware of the critical importance of the Metro programme to BL's future survival. With a Conservative government reluctant to provide further state finance, and Sir Michael Edwardes as chief executive, there was not the remotest prospect of government support being maintained indefinitely.

For these reasons, in the period leading up to the launch of the Metro the Works Committee sought to avoid embroiling the car's production programme in the disputes surrounding the Draft Agreement. Thus in August 1980 Jack Adams - Derek Robinson's successor as plant convenor - was able to say: "Whatever problems we have in the plant, we recognise the priority of the Metro. We have done everything possible to accommodate the Metro" (Interview, 28.8.1980).

New Technology and the Maintenance Trades

Although the effects of the Draft Agreement touched all work groups at Longbridge, the case of the maintenance workers was markedly different from that of the direct production workers. Not only did the maintenance tradesmen remain aloof from the representation provided by the Works Committee, but they also had specific, vested interests to protect in the process of technological change. Moreover, given the more capital-intensive nature of the Metro facilities, the maintenance workers seemed to be the only manual group likely to emerge from that process with their bargaining power enhanced and not diminished.

Management were all too well aware of this possibility, and at the design stage of the New West deliberately sought to minimise the skill requirements of the new BIW technologies. For example, the control and co-ordination of the new facilities was entrusted not to a centralised computer system, but to a network of unsophisticated, solid-state Programmable Controllers (PCs). The breakdown of individual PCs could be easily monitored and rectified, with minimum disruption to the production process as a whole.

However, despite such technological refinements,

the "control of uncertainty" inherent in the maintenance workers' role meant that their behaviour and practices inevitably continued to be of pressing concern to management in the pre-launch period. As with the production workers, the central political problem was one of abolishing the workers' job controls, and replacing them with managerially-defined and therefore "efficient" practices. This was no easy task, however, with the maintenance function based upon seven traditional, solidly organised trades groups - Millwrights, Electricians, Pipefitters, Machine Tool Fitters, Jig and Tool Fitters, Toolroom, and Maintenance Fitters. Management's operational control of these trades was heavily circumscribed, with the maintenance workers in their central, off-line pens refusing to respond to telephones, two-way radios, or even production supervisors. Rigid rules of Byzantine detail governed the allocation of tasks between the trades.

Management aimed to replace this trade-based organisation with one more responsive to the "needs" of automated production systems, and chose as their model the kinds of maintenance practices which Fiat had installed in their advanced "Robogate" BIW shops - Fiat being an appropriate example as much for its industrial relations parallels with BL as for its technological expertise. In aiming for greater flexibility and responsiveness, BL management's proposals involved the erosion of the greater part of the trades' autonomy and the weakening or abolition of many demarcations. However, while management could deal with the direct work-force through the formal, contractual imposition of the Draft Agreement, the political cohesion of the maintenance trades and their scarcity value as skilled men made such a direct attack inadvisable. Instead, management placed great emphasis upon manipulating the technical and political margins of the trades' work, with the technologies of the New West being adroitly exploited as the instruments for such manipulation.

More specifically, the advent of new technologies allowed management to implement a re-distribution of skills in the maintenance area. At the lower levels of skill, a new grade of labour - the "weld setter" - was established to carry out the routine and supposedly menial tasks of tip-changing, electrode dressing and weld setting, which were demanded by the extensive deployment of multi-welder machines in the New West. At the same time, the more abstract electronic and programming skills associated with the computer controlled BIW systems were expropriated for the new managerial "Control Engineering" function. The creation of "Control Engineering" was justified in terms of the higher skills required in the New West - skills not within the existing repertoire of the maintenance trades - but it also provided, for the first time, a "co-ordinating" managerial authority in the maintenance area.

The overall effect of these changes was to establish a hierarchical division of labour in the maintenance function. A division of labour, moreover, with the guiding presence of staff-graded engineers at its apex, and embodying a strong Taylorist tendency towards the separation of manual and mental labour. By these means the maintenance organisation of the New West was enlisted in the service of efficiency, rather than the job rights of the trades groups.

However, it is important to note of management's pursuit of efficiency that Taylorist elements such as hierarchy and specialisation should not be equated with a general de-skilling of the maintenance work-force. Such Taylorist features as were evident were more a matter of breaking down worker job controls - what Littler (1982) terms "bureaucratisation of the shop floor" - than of a simple, unilateral de-skilling process. Although management were able to manipulate the re-distribution of skills in the New West, the latter involved both a de-skilling and a re-skilling of maintenance work: certain trades - in particular the electricians - employed their bargaining leverage to help ensure that at least some of their skills and job controls would be transferred to the new maintenance technologies of VDUs and machine monitoring systems. Indeed, it could be argued that with their greater flexibility and newly acquired electronics expertise the maintenance tradesmen of the New West probably exercised a higher degree of skill than was the case previously.

New Technology and Industrial Relations

That the introduction of advance technologies at Longbridge should have aided the extension and intensification of management's shop-floor control of the labour-force should not in itself be surprising. Historically, for instance, the introduction of the assembly line to Europe was greatly facilitated by the unions' acceptance of it as a "finding of science" (Fridenson, 1978).

At Longbridge too, new technology - either in an ideological or a material sense - proved to be a powerful instrument in management-labour relations and the determination of work practices. On one side, technology was implicitly identified with progress in the Works Committee's "alternative" programme and the unions' calls for investment. On the other, management, by dint of their ability to define the "necessary" practices for the operation of new technology, and by deliberately linking their initiatives to the practical problems raised by technological innovation, were able to introduce radical changes in shop-floor social relations without provoking a head-on clash with powerful interest groups such as the tradesmen. In this enterprise, what was technically necessary, what was taken for granted, and what was simply politically expedient became so intertwined as to be almost inseparable.

Whether one views this political dimension of technology as an important "legitimating principle" (Armstrong, et al., 1981) at workplace level, or as a reference point in the "negotiation of order" (Hyman, 1980), there is little doubt that in the Longbridge case technological change provided a political juncture, a "window of opportunity", in which Taylorist forms of management control could be more firmly embedded in the production process. Although new technology per se was far from being the primary instrument in the re-shaping of work practices - this being performed by the Draft Agreement - it certainly played a useful role in both promulgating and legitimating the re-structuring of production, and in the process unsettling even the most secure union job controls.

Conclusion

The most tangible effect of the period of intense change described above was a dramatic reversal in the productivity trends at Longbridge. Overall productivity at the plant had been in constant decline since the introduction of MDW in the early 1970s. But, in the first full year of Metro production labour productivity increased by no less than 120%: in 1980, 17,000 workers produced 123,000 cars but in 1981 13,000 produced 235,000 cars. The increase in productivity also allowed management to reduce manning levels much more drastically than planned, such that by 1982 the plant's manual work-force, which stood at 20,000 in 1975, numbered only 9,600.

Much the greater part of the increase in productivity seems to have been attributable not to the introduction of new technology but to the imposition of the new managerial regime at the plant; the abolition of mutuality, installation of "tight" work standards, greater labour mobility and flexibility, and so on. Indeed, BL management themselves readily admitted that only 10-15% of the productivity gains derived from the technological change at the plant. This is further underlined by the performance of the New West shop itself, where, despite the fact that 80% of the spot-welding tasks were performed mechanically (against just 10% for the Mini), total man-hours per car were 50% below budget levels, and machine systems designed on the premise of 85% efficiency were achieving 95% or more.

Yet, despite the striking turnaround at Longbridge, the extraction of greater effort and flexibility from the labour force could contribute relatively little to the solution of some of BL's more fundamental and strategic problems. It could not, for instance, achieve greater economies of scale, or the multinational linkages required for full-scale competition with the major producers. Nor could it ensure the correctness of management decisions on marketing strategy, which were equally critical to the firm's future.

In this perspective, the manner of the new technologies' introduction at Longbridge and the autocratic imposition of new work practices seem to have reflected what were essentially short and medium-term managerial imperatives to stabilise the company's finances and enhance the government's perception of its future viability. Such operational changes cannot, however, provide the complete formula for BL's long-term survival. This, it seems, still hinges on political will rather than commercial or financial performance.

REFERENCES

Amstrong, P.J., J.F.B. Goodman, and J.D. Hyman, 1981: Ideology and Shop-floor Industrial Relations, London.

British Leyland, 1975: Employee Participation in Leyland Cars, British Leyland.

Crozier, M. 1964: The Bureaucratic Phenomenon, Chicago.
Edwardes, M. 1983: Back from the Brink, London.
Fridenson, P. 1978: "The coming of the assembly line to Europe", Sociology of the Sciences, vol. ll.
Hyman, R. 1980: "Trade Unions: control and resistance", in: G. Esland and G. Salaman, eds., The Politics of Work and Occupations, Milton Keynes.
Littler, C. 1982: The Development of the Labour Process in Capitalist Societies, London.
Owen, A.E. 1980: "Economic criterion for robot justification", Industrial Robot, vii (3).
Robinson, D. 1980: "Interview with Derek Robinson", Marxism Today, March.
Scarbrough, H. 1982: The Control of Technological Change in the Motor Industry: a case study. Ph.D. thesis, University of Aston at Birmingham.
Team of Inquiry led by Sir Don Ryder, 1975: British Leyland: the next decade, London.

Chapter Six

CHANGES OF INDUSTRIAL RELATIONS AT FIAT

Matteo Rollier

Some Basic Information

Fiat, by far the biggest privately owned company, has always set the pace for the industrial relations in Italy. The development of work relations at Fiat sets the trend for the whole of Italy and the state of power relations at Fiat mirrors the situation all over the country.

Fiat is a multinational company which in 1983 had a worldwide turnover of 15 thousand million dollars and had a workforce of 250,000 of whom about 200,000 were employed in Italy. The most important line of business is the production of cars which amounts to about 50 per cent of the company's turnover. The company is also engaged in the production of commercial vehicles, agricultural machines, aircraft engines, industrial plants, robot controlled manufacturing systems and in other sectors of the market such as telecommunications, electronics, bio-chemistry as well as in the service sector (Comito, 1982).

In 1983 the company had a workforce of 100,000 in its automobile plants, 90,000 of them employed in Italy. The most important foreign location is in Brazil where Fiat's share of the market amounts to just under 15 per cent. In its Italian works the company produces the Lancia as well as the Fiat models. The share of the Italian market amounts to 50 per cent, the share of the European market to 13 per cent. In 1983 the company produced 1.2 million cars in its Italian plants (Streeck/Hoff, 1982).

During the 1970s Fiat faced a severe crisis but has by now successfully carried out a reorganisation programme. The most important measures of that programme were:

Fiat abandoned a great number of plants, e.g.

Fiat withdrew from the steel sector and sold automobile plants in Spain and Latin America.

The company made every effort to raise productivity. As a result, the annual production per worker could be increased from 14 to 24 cars during the last five years. Fiat has above all modernised the technological standard of its manufacturing plants. By the end of 1984 900 industrial robots will be installed in its automobile factories. According to Fiat statements, Fiat is the most highly automated car producer in the world.

By selling plants and by dismissals, Fiat has drastically lowered the level of employment. In the last five years the company as a whole has shed more than 100,000 employees; in the Italian automobile plants the number of employees has been reduced from 140,000 to 90,000.

An essential part of the reorganisation programme was to regain the predominance over the labour market and the work organisation which had been lost since the "autunno caldo" at the end of the sixties to the trade unions. An important turning-point for the company was the notorious "march of the 40,000" in 1980 which was staged with the support of the company and which was a demonstration against the trade unions by foremen, supervisors and white-collar workers such as clerks and technicians (Baldissera, 1983; Mattina, 1981).

Turning-Point of Industrial Relations

In June 1980 Fiat announced the urgent need for an immediate 30% cut in production, and, as a result, a considerable reduction of staff. After a long and dramatic struggle, an agreement was signed at the Employment Ministry on 14th October. This agreement excluded the resort to widespread dismissals but also placed 23,000 workers in the "Special Redundancy Fund Scheme" (Cassa Integrazione Speciale) up until 31.12.81. The "Cassa Integrazione" provides 80% of the average salary of workers, who formally remain occupied by the factory; at present there is no time limit to that redundancy fund. Along with the "Cassa Integrazione" scheme, other measures included in the agreement were to reduce staff consisted of "incentives for voluntary resignations", "early retirement schemes", and "internal mobility within the group". The agreement also established that, before 30th June 1981, an investigation would be carried out to see if any excess employment could still be found after the above-mentioned measures had been exploited: if so, the agreement provided for "external mobility, from one workplace to another". Subject to the carrying out of all these measures and checks, Fiat committed itself to finding jobs for all those workers who were still in the "Cassa Integrazione" scheme by June 1983.

Since then part of these measures have been fulfilled, above all the voluntary resignations. External "job-to-job" mobility has remained an exception, while the "Cassa Integrazione" scheme has been extended beyond even the final date set down in the agreement - this meant that another agreement had to be signed in October 1983 which, with 15,000 people still in the "Cassa Integrazione" scheme, provides for further partial re-employment and checks until 1986, thus leaving the problem unresolved for about 11,000 people.

In the meantime about 17,000 people have handed in their notice at Fiat (1982 figures), of whom less than half were "Cassa Integrazione" workers. This means that many workers took advantage of the good terms of resignation offered by the agreement; but it also means that the workers who remain in the "Cassa Integrazione" scheme have very real and serious problems in re-integrating themselves in the labour market.

The 1980 dispute indicates a turning-point in industrial relations, and not just at Fiat. Historically it follows the period of "national solidarity" (1977/78). At this time the Communist Party formed part of the parliamentary majority supporting the government and the unions had tried out a substantially "neo-corporatist" project in which they offered wage moderation in exchange for economic policy and employment protection measures. Since the entrepreneurs had not taken much action in this period, the Fiat dispute marks the turning-point and indicates a vigorous return to action and an overall hardening-up of industrial relations (Dal Co/Rollier, 1984).

There is no doubt that the agreement enabled Fiat to avoid resorting to direct mass dismissals by introducing a cushioning mechanism. By the end of the process initiated by the agreement, however, the reduction of staff will be quite close to that requested by Fiat at the beginning. Furthermore, for long periods such as those considered here, the "Cassa Integrazione" scheme produces devastating social effects as well as serious consequences on the labour market.

The bitterness of the dispute and the arguments which accompanied it, the divisions among the workers, produced by the management of dispute and, above all, by the split between those who kept their jobs and those who were placed in the "Cassa Integrazione" scheme (the union request for a rotation scheme was not accepted): all these factors have produced deep wounds which cannot be healed either easily or quickly. Even worse, a general climate of frustration, lack of faith and fear has resulted which currently seems to be blocking any possibility of union action, even in more traditional and everyday areas.

The process of staff reduction has been accompanied by profound technological and organisational transformation. The mechanism of the 1980 union agreement obscured the question whether the declared employment excess was technological in origins or due to market problems. Later checks did not resolve the problem. In retrospect the second hypothesis seems more plausible at least as far as the original situation was concerned. However the determination with which the problem of staff excess was posed, as a whole and immediately, thus rejecting from the very beginning more gradual strategies which would have obtained the same results without provoking such a dramatic collision - all this leaves no doubt that the Fiat management believed that it could thereby provoke a "show-down". This would free it from the internal system of industrial relations which was consolidated after the dramatic events of autumn 1969 and remove the obstacles to technological and organisational restructuring on a grand scale. Now that part of this restructuring has been achieved, the excess of staff is probably no longer cyclical or due to market conditions but structural and technological in nature. But this change took place <u>during</u> the process initiated by the 1980 union agreement. Moreover the complex mechanism of the agreement has meanwhile brought about a "<u>reselection</u>" of the workforce which is a function of the technological and organisational evolution itself.

Union Strategy on Working Conditions

From the end of 1968 Fiat had been, if not the most advanced then at least the most important, hot-point in that particular climate of industrial relations which characterised Italy in the seventies. During the dramatic autumn of 1969 it had been the stage for a great explosion of working-class conflict. This was built up by the unions but exploded largely outside their control. They were then able to lead it back to a fundamentally important experience of democracy and participation: the setting-up of "consiglidei delegati" (workers' delegates committees) and the 1969 and 1971 agreements which guaranteed delegates with important control and intervention rights about work organisation.

The unions had previously been very weak in the Fiat group, but in those years it reached levels of union membership close to the average in Italian industry. It has been observed that, intervening in areas of labour organisation and making use of these new forms of representative organisation, the unions in these years were able to extend the area they represented (which up until then had been small-scale not just at Fiat), to less specialised workers, assembly line workers, and immigrants from the South (Pizzorno, 1978). In the light of subsequent events it should be added that the relationship at Fiat between these workers and the union has always remained critical: delegating to the union has probably always been something quite temporary and "instrumental" and has never been transformed into something more permanent and consolidated. If this were not the case we cannot explain the rapid decline in strife after the crisis in industrial relations at the end of 1980, something which has no parallel in other situations where the attack on employment levels was just as head-on and dramatic. It should be said, however, that union strategy to control and intervene in labour organisation was already in widespread crisis before 1980, and this would better explain what happened afterwards.

We can summarise Italian union strategy towards labour organisation from 1969 onwards as the progressive construction (by means of company-level bargaining) of direct instruments (entrusted to the delegates) of control over those aspects of labour organisation which most directly influence working conditions such as: assigning work tasks, loads and speed, staff, internal mobility, specialisation, work environment. Such control was carried out through the right to know the regulations (for example the regulations for assigning work hours and deciding assembly line rhythms in relation to the number of workers) and the right to guarantee respect for the regulations in the case of variations (this right was placed directly in the hands of the delegates). Though these rights may seem quite basic and rudimentary, their importance can only be appreciated if we recall the extremely authoritarian character of traditional labour organisation, in particular at Fiat.

In many cases, and especially at the beginning, the organisational regulations were those already established: that is to say, the unions did not question the regulations which already existed and which had, obviously, been established by the company. In other cases supplementary regulations were gradually introduced, as in the case of the 1971 Fiat agreement which secured, for example, a rigid, inverse relationship between assembly line speeds and saturation maximums. In such cases the application of such regulations was handed over, once again, to the delegates. Important supplementary regulations were introduced in the areas of work environment and work categories. The "single grading system" for white and blue collar workers introduced by the 1973 National Metallurgical Workers Contract, for example, involved a period of application at the level of the firm which implied re-appraising the whole system of categories at this level. Once again this was to be negotiated by the delegates.

There is no doubt that one result of all this was an overall stiffening up of work organisation. This was true in two senses: in general, because every decision which modified any of the organisational factors had to be discussed with the delegates concerned, and naturally enough these discussions were not limited to purely technical aspects; and, in particular, because it precluded an elastic use of organisational regulations by managerial staff - something which had been the condition for survival of the Taylor-Ford system of organisation for many years.

In passing we can note that this union strategy has important analogies with the rigid definition of work-rules carried out by many union movements much earlier elsewhere than in Italy - as a protection against the spread of "scientific-management". It should be pointed out, however, that in most cases the work-rules are limited to regulating the residual areas of labour organisation (or removing them from the discretionary power of management). In this case, however, the system directly provides for the day-to-day exercise of fundamental managerial prerogatives.

This "hardening up" of labour organisation was intentional as far as the unions were concerned, but it was not regarded as an end in itself.

Successfully combining the control exercised by the delegates at shopfloor level and company bargaining carried out by the "consiglidei delegati", the union was aiming at:

a) effective day-to-day protection of working conditions (a protection which can be defined as "static").

b) the growing definition of forms of labour organisation which, while increasing production both quantitatively and qualitatively, enabled the quality of working life to be improved. This explains the Italian unions' interest in new forms of labour organisation which could provide "dynamic protection". This inevitably shifts attention away from the rejection and deliberate undermining of the traditional Taylorite system of labour organisation.

c) the growing definition of alternative "decision-making models" which affect not only the overall objectives of the company but also its day-to-day running. Indeed there is no doubt that implicit in these strategies was the objective of "industrial democracy", a level of "participation" hitherto unrealised.

This model of action was undoubtedly effective not only for the day-to-day protection of working conditions but also as an incentive towards organisational innovation. This applied particularly to a company such as Fiat, where the continuation of centralised and hierarchical-authoritarian structures and procedures was especially marked and where, at the beginning, of the seventies, there was still unquestioning faith in Taylorism.

Transformation of Labour Relations and Re-Selection of the Work Force

As soon as the situation began to change, however, first of all through the pressure of union action, and then because of the growing emergence of dramatic market problems, this model of action proved ineffective for organising in a period of transformation. It thereby revealed all the contradictions which had not been resolved.

In order to be effective in a situation of rapid transformation, the union model of action described above should not have been used in a purely defensive way, with the sole aim of turning power relations to its own advantage. Instead it should have been used as a "management" instrument with the aim of directing the process of change and consolidating results in terms (even if partial) of new forms of labour organisation and new decision-making models. All this would have required regulating rigidity on the basis of the intermediate results of a strategy.

It does not seem that this was the case. For example, when intervention and control measures taken by the painting-room workers led the company to introduce substantial technological changes which markedly improved working conditions, the unions and the delegates were unable to give sufficient guidance to the workers that such changes meant that the previous allowance due to the old, very bad working conditions was no longer justified. As a result there was a battle to keep the allowance which, independently of the compromise finally obtained, had the result of discrediting, in the eyes of the workers, the achievement of an important objective in the area of labour organisation. A similar story took place with the March 1974 union agreement. This introduced important work experiments no longer based on the assembly line, namely, the "assembly islands" (semi-autonomous work groups for engine assembly and also, experimentally, for car-body assembly). But the unions could not follow the introduction of the changes step-by-step, discussing possible solutions and proposing possible changes. Thus, with the production crisis which followed the first oil "shock" in 1974, these experiments were partly abandoned and partly "normalised". In the final analysis the experience benefited only the management who were able to use it profitably for their future programmes.

These limitations probably depend above all on the intrinsic weakness of worker-union relations at Fiat; but there is quite certainly an underlying inability of the unions to take on all the "management" implications of the union model of action.

The union model of action has been applied in a fundamentally ambiguous way and has therefore been limited to an exclusively defensive role. The idea of changes in labour organisation has only been used in terms of "agitation" since this was the area felt most strongly by workers, the area in which it was possible to direct the strife without, however, building up a "strategy of organisational change" by means of these periods of conflict. A centralised logic of "political exchange" prevailed in the period of "national solidarity". This meant not bringing too much pressure to bear in this area which, as we have said, directly involved fundamental managerial prerogatives. On the other hand the "management" implications of this model of union action have never fully convinced an important part of the union movement.

In the meantime the union was necessarily clarifying its general strategies in order to face up to the crisis and problems of employment: these included industrial policy strategies and strategies for the protection of employment. But what was lacking was the working out of a policy which inserted organisational change within these more general strategies. As a result union action about labour organisation has remained isolated and separated from the general union strategies.

Yet when, in the face of a situation which was no longer static but moving rapidly, extensive processes of change were set underway, this "rigidity" strategy became ineffective even for exclusively defensive aims.

The unions had already found themselves in difficulty, during the first oil crisis, when they were unable to control the phenomena of internal mobility and the rapid changes in organisational parameters (production quantities, work rhythms, staff).

So when automation and computerisation processes began to develop alongside the introduction of new models (following the new car "project philosophies" developed as a consequence of the oil crisis), the control instruments which had been constructed on the basis of traditional labour organisation were found to be completely ineffective.

In many cases these instruments have been quite simply dismantled by the other side; but in many other cases, and above all at Fiat, they have remained in operation. Delegates still control their own work teams and in certain areas are able to impose some forms of "rigidity" but these are no more than the ruins of a building which no longer exists and are quite useless in protecting workers from the possible negative consequences of restructuring and even more useless in influencing the restructuring itself.

In this way both the labour organisation strategy and the general union strategies are separated from each other, and in workers' eyes they seem less and less credible - this has set off a legitimation crisis of the union which is particularly dramatic at Fiat today. There is no doubt that this crisis is the result of the very violent counter-attack undertaken by management and marked, as we have already said, by the late-1980 dispute; but it is also a consequence of the lack of solution to the problems analysed here.

In order to complete this analysis of the underlying reasons for the present crisis in industrial relations at Fiat, it should be added that shortly before the explosion of the crisis in 1980 more than 10,000 workers had been taken on (and this in a period of extreme instability of market trends for Fiat), apparently in order to face up to urgent production requirements which, with the rigidity of the unions over internal mobility and working hours, could not be resolved in any other way.

This, too, is indicative of an overall climate in which any dialogue was totally lacking. In the light of events only a few months later, this decision on the part of management seems at least somewhat hazardous. But what interests us here is that this massive introduction of young workers, who arrived with a culture and patterns of behaviour completely different from those of the traditional working class, increased (at the very least) the unions' problems of legitimation and their ability to guide the workers.

In fact this large mass of young workers found that it could use, though only for a short period, those instruments for protection of the working-class, those guarantees and rights which had previously been won, but within a system of values and individual and collective behaviour which was completely different from their own, and without realising, what is more, the extreme fragility of such conquests.

Some observers have seen in this the emergence of "new working-class behaviour". They view practices positively which use collective guarantees for individual self-protection. The majority of these young workers are today amongst those in the "Cassa Integrazione" scheme as well as amongst those who will find it more difficult to get back their jobs in the factory. But it seems quite difficult to view such behaviour today (by now it has completely disappeared) as little more than contingent. The fact remains that this factor has further damaged the unions' means of intervention, as well as further undermining their ability to represent the workers.

All these factors together should thus explain, first of all, why, at the end of 1980, the Fiat management saw the opportunity and possibility of accelerating a bitter, head-on conflict with the unions, and why they had a very good probability of winning it since the advantages to be had were greater than the predictable costs resulting from the destruction of the union side of the dialogue.

Secondly, they should also explain - apart from any consideration of the way in which the dispute was carried out - why the union was not able to develop a solution which, while reducing the number and length of time of work suspensions as far as possible, would allow the maximum control over the relationship between employment and the processes of technological and productive restructuring.

The outcome of the 1980 dispute represented an important factor in limiting the attack on employment levels but it also entailed losing a degree of control and intervention over labour organisation. This confirmed the rupture between (a) what the delegates might still be able to do to control the technological and organisational restructuring on the shopfloor in order to defend working conditions, and (b) the examinations carried out at centralised union level to verify the productive economic trends of the firm and overall employment.

It is clearly not the job of a study such as this to evaluate the possibilities of reaching a different solution along these lines. The analysis so far, however, would suggest that the limits that weighed on the dispute had older and more complex origins. The fact remains that the agreement opened up a new phase in industrial relations at Fiat, and that the "model of action" - so well-known and widely discussed, even abroad - of the union at Fiat, was completely overthrown as a result.

As a result of the events surrounding the 1980 agreement and the investigations which followed it, the process of technological and organisational restructuring continued and, to a large degree, <u>outside</u> any union control.

Before describing this process, however, and its consequences for industrial relations, it is important to analyse the process of re-selection of the labour force which resulted from the work suspensions carried out after the agreement. A sample analysis carried out in 1982 proposes a very interesting account of this process (ISFOL, 1983). As a result of the "Cassa Integrazione" mechanism and the internal mobility which accompanies it, what is happening can be summarised as follows:

(a) a radical expulsion of the weaker sectors of the labour force (female, young, old and handicapped workers);

(b) a "re-selection" of the "central" core of the labour force, with the tendency to eliminate those groups which are presumably less integrated into the company system and the social context.

With this mechanism of a progressive reduction of employment, and considering the large number of workers taken on previously, one might think that the company intended to carry out a forced and accelerated employment turnover with the aim of rejuvenating the workforce. But ISFOL's study shows that exactly the opposite is true: the large majority of newly-hired workers lost their jobs, while those who have kept them, or will do so in the future, seem to be workers who are thoroughly settled and stabilised, both from the productive and the relational point of view, with medium-to-long term seniority in the firm, a relatively high level of specialisation, and a normal, settled family and emotional situation.

The exact opposite, that is, of the "mass-worker", the man on the assembly lines, the bearer of values and attitudes, the worker who, in the case of Italy, was typically young, newly immigrated, single and without any strong ties in the city area, and not at all used to working in a large firm.

So it should be admitted, if this analysis is correct, that a kind of social revolution is underway within the Fiat complex. Naturally enough it is impossible to say what aims management had in mind when they set it underway - it could even be maintained that, since management has to reduce staff, they decided that it would be less dramatic if they first dismissed those who could, at least theoretically, find another job or move back into the family. But the fact remains that the social and cultural universe which had been the expression for the previous struggles, was definitively overturned and, whether intentionally or by coincidence, the workers who remained were socially and professionally perfectly suitable for the new technological and production conditions of the firm.

Technological and Organisational Transformation

It would of course be schematic and misleading to argue that all this transformation occurred in the period after 1980. Some important processes were set underway much earlier: first of all, the process of splitting the Fiat complex into divisions, the introduction of more decentralised and flexible managerial structures, together with a series of measures to improve the economic and financial situation, and secondly, a rationalisation of the "make or buy" policy, and of the networks of suppliers.

Within these two policy lines, the process of technological and organisational evolution gradually accelerated and extended after the first production crisis after 1974 and the ensuing period of uncertainty about the basic strategies of the company.

The early technical and organisational measures seem to be a reply to immediate requirements, which often involved answering union requests or eliminating sources of strife (which, as we have seen, the unions were not able to "organise" strategically on their own). Later it would seem that management gradually clarified an overall "philosophy" which tied the product strategies to strategies of technological and organisational choice and linked the latter to an idea about internal social regulation.

As a result, while the early measures did not change the basic organisational schemes (which, simplifying, can be called "Tayloristic") and were limited to mechanising certain work stages, and introducing buffers where the chain of operations was too rigid, afterwards they appear to be new attempts, the basic "philosophy" of which, however, seems to be uncertain.

This is well illustrated by the already-mentioned setting-up of "assembly islands" for engines and car bodies. This had been discussed with the union throughout 1973 and, as has been said, was the subject of the March 1974 agreement, being presented with strong social emphasis and participation. It is also illustrated by the setting-up at almost the same time of "Digitron", a system of semi-automatic coupling of car bodies and engine parts with fixed work stations and work groups.

The setting-up of "Robogate" seems to be a turning-point in the project philosophies of the firm. This is nothing else than a system of welding robots integrated by a handling system of grid pieces and run by a computer. It represents a complete transformation, not just of the machines, but also of the overall organisation around them.

This philosophy seems to be based on a precise hypothesis: that the potential offered by technology makes it possible to resolve, not just problems of production flexibility and productivity, but also the majority of social problems and the sources of strife. For this reason, the problem of technological and organisational programming needs to be restored to its exclusively technical aspects, removing it from the sphere of social problems and, above all, from the sphere of industrial relations. Whether this hypothesis is founded or not, and whether it can really offer a stable solution to social problems and an effective use of the new technologies, is something we cannot tackle here. None the less this writer doubts it very strongly. The fact remains that this seems to be the path chosen by Fiat management.

On this basis, and parallel to the attack on the area of industrial relations, it was possible to take away a large part of the union power of intervention in labour organisation. It was possible to programme and organise enormous technological and organisational restructuring, outside of any union participation and control.

Together with this system - and many other minor, less noticeable innovations in the machinery, presses and painting process - another system was then programmed and set underway. This was the LAM (Asynchronous Engine Working), a system of engine assembly with fixed, individual work positions operating over all the work stages, with a handling system of grid pieces on self-propelled trollies, and run by a computer. Finally, at about the same time, completely automatic assembly lines were being set up for gears and engines.

In this way Fiat has developed from an archaic firm on the level of managerial techniques, but "mature" on the level of productive technologies, to a markedly innovative company on the level of organisation, technology and product policies.

This poses new problems. We can cite an article by a Fiat manager which very clearly describes these new problems:

> Bureaucratically and rigidly established forms become less indispensible when the role of control is, in large part, handed over to the systems of information. In brief, the flexibility of the process increases, but its level of internal integration increases considerably as well. The evolution is thus towards forms of labour which are less parcelled out, less tied down, displaying features of greater integration and directed towards a technical power of guiding the process which is based on adequate knowledge and information (Besusso, 1983; the underlining is ours).

As a result, this problem is very important. Both the organisation of the technological and organisational transformation and the day-to-day running of the advanced technologies require a high level of worker trustworthiness, a certain level of involvement, and a certain capacity for "innovation", all things which are completely foreign to the traditional Tayloristic form of organisation. The re-selection of the labour force mentioned above is useful in fulfilling this requirement; but it is not, of course, enough in itself over a medium-term period. For the problem then arises of what kind of social relations, and thus industrial relations, Fiat wishes to build up in the future.

The trends which can be picked out emerge from two facts which can be analysed as follows:

1) Resorting to incentives which do not so much reward the quantity of production as precisely this involvement by workers in the problems of running the technological innovation. This "involvement" is not necessarily a "skill" (even if it can perhaps be developed in that sense) and even less does it represent a form of "participation".

One example of this is the premium on LAM fixed by the unions in 1983. To confirm the importance of this fact, it is essential first of all to note precisely that this premium was fixed by the unions in negotiations only with the delegates within the departments, and this represents the only case in which Fiat not only accepted but asked to negotiate something which concerned the organisation of labour after the events of 1980.

The premium rewards the reduction in unworked time due to the bad working of the handling system (which is very complex, given that the machinery uses many types of motors). The reduction in unworked time depends on the speedy measures taken by workers to bring the system back to normal, since otherwise the errors multiply exponentially. It is clear that when something went wrong in the traditional Tayloristic organisation, the worker was not asked to do anything: today, on the other hand, he needs to take action to a certain degree. Thus the premium represents a first general policy line by Fiat management to respond to the problem of "involvement". The fact that the premium was discussed with the union within the factory demonstrates that, within these limits, and on the condition that they are not exceeded, the union is recovering its role of dialogue with management. We could say that Fiat is trying to construct an interlocutor suitable for the new rules of the game.

2) Experimental forms of direct involvement which by-pass the collective mediation of the interests represented by the unions. This is above all a question of management style and climate: as a result the task is handed over to the foremen, within the framework of redefining their role. This need became urgent with the well-known "foremen's crisis" and the famous "march of the forty thousand". But it is also an objective which is being sought by means of the cautious, but more and more widespread, introduction of discussion groups, or by solutions such as "Quality Circles".

This would seem to be a point of central importance, and not just at Fiat. It appears to be the most interesting reply to the requirement for "involvement" which we spoke about earlier. A criticism of this solution lies outside the province of this paper, and the experience at Fiat is anyway too recent for us to be able to draw conclusions of any sort. The fact remains that, if this policy line were successful, it would deeply modify the context of industrial relations, at Fiat as elsewhere. It is difficult to predict in which direction: it seems probable that, in a context which is very different from the original Japanese one, such forms will give rise to completely different phenomena. And the fact that these forms are being introduced today after a period of crisis in the unions, after a period of defeats and concessionary bargaining (with the result that such forms would seem to be a substitute for work-rules, bargaining and participation), could be irrelevant in the long run. It should be recognised, however, that in the short run these forms become part of the limits of union action, and find by no means marginal agreement among the workers. This is something very similar to what other authors in this research have called "neo-populism" by which the supporters of free-trade theories manage all the same to obtain mass consensus.

On the other hand, the technological and production restructuring already carried out permits an important recovery of productivity in certain sectors and firms, especially when symptoms of general recovery are to be found. This creates a certain space for wage concessions which, however, are strictly tied to productivity and the development of an attitude of participation and involvement which we spoke about earlier. Today this probably interests management much more than the definition of centralised mechanisms of wage regulation and other matters involved in industrial relations. In other words, on the basis of what we have said here regarding Fiat, we foresee, not so much an end to bargaining at company level, as the redirection of bargaining towards different problems and according to very different rules of the game.

Caught between a crisis of representation determined by the transformation of the firm's social system, and the loss of effectiveness of its own systems of intervention and control over the production processes, and trapped in a hostile spiral triggered off by the separation between its own general strategies and its action on the shopfloor, the union seems to have no other alternative today than that of joining in these new rules of the game, the outcome of which, however, is by no means decided. It is impossible to predict what will happen: the aim of this paper, however, was to account for the end of a "paradigm" of industrial relations of which the situation at Fiat was the most typical example, and the emergence of a new, though still uncertain, new "paradigm" with respect to which, however, the Fiat complex is still confirmed to be an advanced laboratory of Italian industrial relations.

REFERENCES

Baldissera, A. 1983: La marcia dei quarantamila: una critica delle interpretazioni correnti, in: chiesi, A.M. ed. Le ricerche sui lavoratori non manuali e il sindacato in Italia, Milano.

Besusso, C. 1983: Evoluzione tecnologica e organizzazione del lavoro alla Fiat, in: Produrre, 10-11, 1983.

Comito, V. 1982: La Fiat tra crisi e ristrutturazione, Roma.

Dal Co, M. e Rollier, M. 1984: La crisi delle relazioni industriali in Italia, in: Perulli, P and Trentin, B. ed., Il sindacato nella recessione, Bari.

ISFOL, 1983: Caratteristiche e comportamenti degli operai Fiat in mobilità, Quaderni ISFOL, 3/1983.

Mattina, E. 1981: Fiat e sindacati negli anni '80, Milano.

Pizzorno, A. 1978: Lotte operaie e sindacato: il ciclo 1968-1972 in Italia, Bari.

Streeck, W. und Hoff, A. (eds.), 1982: Industrial Relations in the World Automobile Industry - The Experiences of the 1970s, Science Centre, Berlin.

Part Three

CRISIS AND RATIONALISATION: IMPACT ON UNIONS

Chapter Seven

BUREAUCRACY, OLIGARCHY, AND INCORPORATION IN SHOP STEWARD ORGANISATIONS IN THE 1980s

Eric Batstone

The debate on what can very loosely be termed "corporatism" has placed considerable emphasis upon the characteristics of trade unions at the national level, but has tended to ignore the role of workplace organisation (for exceptions, see Pizzorno, 1978; Sabel, 1983). Even when we look at the more highly corporatist societies, however, we need to take fuller account of the collective organisation at the place of work and how it relates to the larger union structure. This is so for at least two reason; first, it is clear that in a number of "corporatist" societies collective organisations at plant and company level enjoy considerable - if constrained - discretion. This is indicated, for example, by the level of wage drift: between 1953 and 1965, drift accounted for 52 per cent of the increase in earnings in Sweden, and for 22 per cent in Austria - both "corporatist" societies - compared with only 19.5 per cent in Britain (UN; 1967: Ch. 3, pp. 32-3). Second, the Swedish experience suggests that, where attempts to reduce workplace discretion are made, the corporatist system may be subjected to severe challenges (Hart & van Otter, 1976). The role and nature of workplace organisations is therefore crucial: it would appear that the conditions for stable "corporatist" arrangements include some degree of autonomy for workplace organisations; but, at the same time, autonomy has to be buttressed and constrained by the state and/or the national unions; and, finally, the extent to which the workplace organisations can directly shape union policy at national level may have to be limited (for a fuller discussion, see Batstone, 1984, Ch. IX; see also Korpi, 1978, and Streeck, 1982).

This weakness in the debate on "corporatism" has

been less serious in the analysis of British experience, if only because in large parts of British industry the workplace is so evidently the locus of union power. In the 1970s, however, some writers suggested that a two-fold strategy was being implemented to provide the conditions for corporatism (e.g. Hyman, 1979). The first strategy was the incorporation of shop steward leaders into the larger unions, with the result that these stewards would induce support on the shop floor for corporatism. The second strategy was a form of enterprise corporatism. This involved the reform of workplace industrial relations on the Donovan model, and subsequently, the introduction of various forms of union-based consultation. Plant and company level reforms of this kind would create a management-sponsored, elitist and bureaucratic shop steward organisation which would adopt a managerial perspective, facilitating greater efficiency and worker quiescence. Some - at least prior to the demise of the Social Contract - claimed not only that this two-pronged strategy was being implemented, but that it had already met with considerable success.

However, this thesis received little factual support. While it is true that during the 1970s a number of unions gave fuller recognition to shop stewards and workplace organisations, there is little evidence that this led to their subordination by national union leaders. Indeed, the reverse tended to be the case. For workplace representatives used their new-found power not only at local level but also at national level to impose constraints upon union policy (see e.g., England, 1981; Undy et al., 1981; and Batstone, Ferner and Terry, 1984). Similarly, there is little evidence to suggest that steward incorporation at plant or company level met with any significant degree of success. It is on this latter theme and its implications for events over the last five years that this paper will focus.

The Decade after Donovan

The thesis of steward incorporation, or the "bureaucratisation of the rank-and-file", argued that Donovan-type reform was aimed at reducing steward and worker power by developing a managerialist and oligarchical steward body which would be prepared both to sign workplace agreements (which would shift the frontier of control in management's favour) and to ensure that workers in practice conformed to the terms of those agreements. Crucial to this strategy, it was claimed, was the development of a hierarchy in domestic organisations - the growth of senior stewards, regular shop steward meetings and full-time stewards - along with a growing dependence of shop stewards and trade unions upon resources, such as check-off, time-off and office facilities, provided by management.

While it would be wrong to attribute the changes primarily to a widespread espousal of the Donovan philosophy (more important were the growth of personnel specialists associated with a more general increase in the management division of labour due to increasing industrial concentration, and the need to protect companies from industrial relations legislation), there certainly were significant changes in patterns of workplace organisation from the late 1960s. The number of stewards increased dramatically, as they spread into new sectors and companies and as steward density increased; more plants had senior stewards, regular stewards meetings and full-time shop stewards; the provision of various facilities for unions and stewards increased; and, although it is not possible to trace it statistically, it is generally accepted that plant bargaining (and later company bargaining) became increasingly common. Along with these changes, there were widespread substantive changes, in the form of productivity agreements, changes in payment systems, and the introduction of such techniques as job evaluation and work study. Later, various forms of "participation" were widely introduced.

But how far these developments actually entailed the development of an incorporated and oligarchical pattern of steward bureaucracy is open to doubt. This is so for a number of reasons. First, the thesis rests upon an inaccurate model of shop steward organisation prior to Donovan: even prior to reform the typical shop steward was - in the oft-quoted phrase - "a lubricant rather than an irritant", and, where workplace organisation existed, some form of hierarchy was common. Second, while there has been an increase in the <u>number</u> of shop steward bodies which have "bureaucratic" features, it does not follow that there has been a corresponding increase in the <u>proportion</u> of all steward bodies which have these characteristics. In fact, only a small minority of shop steward organisations in private manufacturing industry in

1978 had senior stewards <u>and</u> regular steward meetings, and even fewer had <u>full</u>-time shop stewards as well. Third, the available evidence does not support the view that these "bureaucratic" features were generally associated with an oligarchic and incorporated steward leadership: there is no systematic evidence supporting contentions that steward or senior steward tenure had increased, that there had been an absolute centralisation of bargaining, that steward influence and control over members had increased, or that shop-floor bargaining had declined. Further, there appeared to have been a dramatic increase in the range of issues negotiated and the amount of bargaining not only for senior stewards but also for "rank-and-file" stewards. Fourth, the available evidence suggests that, far from reducing steward influence, substantive reforms often increased it: agreements embodied and extended worker and union rights; the new rules were mutually contradictory; and workers and stewards increasingly bargained over the implementation of the jointly agreed rules (in contrast, custom and practice had very often worked in management's favour). Fifth, despite the increasingly adverse economic climate, the "outcomes" of industrial relations practice - in the form of strikes, productivity growth, real earnings, job loss relative to output, the share of profits and wages in GDP - indicate that, in the period 1969-78 compared with the previous decade, workers rather than management gained from reformism. This is true even in those sectors where shop steward organisation was relatively new. (I have discussed these points at length elsewhere, e.g. Batstone, 1981; and especially Batstone, 1984). Finally, it is worth asking why - if the incorporation strategy had been so successful - employers felt it necessary to introduce participation schemes from the mid-1970s.

In brief, while shop steward organisation had grown significantly in the decade after Donovan, and while some "bureaucratic" features were to be found in a large number of plants, shop stewards did not differ dramatically in their essential characteristics in 1978 as compared with 1968. Shop stewards and workers still enjoyed considerable, and probably greater, control over the organisation of work and were able to obstruct and exploit the operation of "reformed" workplace arrangements. But, equally, that meant that steward organisations typically suffered the same weaknesses as they had prior to Donovan: they still tended to be fractionalised (although possibly rather less than in the past) and found difficulty in co-ordinating strategy and action, particularly across plants. This was an especially acute problem given the increased concentration of British industry and the tendency towards the centralisation of strategic decision-making (although not bargaining) within management.

The present state of the union movement stems, to a significant degree, from these developments in shop steward organisation. This is so for three, interrelated reasons. First, they meant that during the 1970s British trade union structure became less and less compatible with the development and pursuit of a coherent political strategy. Such a role as the TUC was able to play at national level stemmed from state sponsorship as much as from any membership mandate. The trend towards a bifurcation of power within the union movement accelerated and the net result of attempts to "negotiate internal order" tended to favour shop steward and union sectionalism. The nature of union-Labour Party links was also important here. The constitutional domination of the latter by the former leads to volatility in Labour Party policy between periods of opposition and office. In opposition, the Labour Party depends upon the support, or at least acquiescence, of the unions so that it may espouse policies with which the party leadership has little sympathy. In power, the Labour Party leadership finds itself less constrained by union forces, as it enjoys electoral legitimacy and confronts other pressures. Labour Governments - as in the 1970s - tend to exploit this freedom, particularly given the traditionally instrumental and often almost casual approach of the unions towards political issues (e.g. Currie, 1979). In turn, this change of emphasis on the part of the Labour Party in power can encourage the traditional union hesitancy over political strategies - and this tendency was exacerbated in the 1970s by the growing influence of shop stewards both within industry and within the unions' corridors of power.

Second, not only did the growth of shop steward organisation impose severe constraints upon how far the union movement could adopt a coherent and committed political strategy, it also played a not insignificant role in the declining international competitiveness of the British economy. This, in turn, exacerbated the problems of developing and maintaining corporatist policies at national level.

Workers' expectations were frustrated, leading finally to the demise of the Social Contract. And of equal importance, the short-term and sectional successes of shop stewards weakened their own position in the larger term. There was a limit to how long unprofitable activities could be continued or companies would continue to operate in Britain when they could produce more profitably abroad.

Third, there were electoral implications of the growth of shop stewards and the related failure of corporatism. Here one can only speculate (the electoral implications of "corporatism" have received rather scant attention), but a number of points seem worthy of consideration. The spread of reformism and shop steward organisation and the claims of the Social Contract increased worker/union member expectations at the same time as they failed to meet those expectations. Moreover, it seems that the attitudes of union members were ambivalent: their essentially economic demands induced union leaders into close dealings with the state; but, particularly when the results were so limited, such dealings were seen to contradict the local democratic traditions in which British trade unionism is firmly rooted (Kahn-Freund, 1979). Finally, it became increasingly obvious - as groups of workers challenged the Social Contract - that its underlying logic had only limited validity. Thus the Conservatives could not only point to the failures of "corporatism", but could also tap the underlying traditions of local union democracy as well as the economistic desires of union members.

The result, therefore, was a Conservative Government, committed to introducing legal constraints upon the unions, ending of intimate government-union dealings, and, most importantly, a monetarist policy which could only operate if union power was drastically reduced and/or if worker attitudes and behaviour were dramatically transformed. These policies, along with other factors, exacerbated Britain's competitive position in the context of world recession as relative unit labour costs rose from 96.2 in 1978 to 147.2 in the third quarter of 1982. But what is important for present purposes is that the role which shop steward organisation played in the 1970s was a factor in its own weakening. For it did not facilitate the development and implementation of coherent political strategies, on the one hand, and, on the other, it played a role in accelerating the long-term trend of falling British competitiveness.

Economic Trends Over the Last Five Years

The results of the failure of "corporatism" are well known. According to official statistics, unemployment rose from 5.7 per cent in 1978 to 12.7 per cent in July 1983; employment in manufacturing fell from over 7.1 million in June 1978 to less than 5.4 million in June 1983 - a reduction of 25 per cent; and union membership fell from about 12.2 million at the end of 1978 to 10.5 million at the end of 1982 - a reduction of 14 per cent. Given these facts, and the well-publicised experience of a number of companies, it is often argued that something approaching a decimation of shop steward organisation has occurred. The very fact that employment had fallen so dramatically in manufacturing - faster than the fall in output - obviously indicates that shop stewards and trade unions are not all powerful. But few claimed that they ever were. The failure to defend employment more successfully is quite predictable given the inability to develop and maintain meaningful relationships with the state in the earlier period and the traditional, sectional approach of shop stewards.

It does not follow, however, that shop stewards have lost all power. Indeed, official statistics suggest that they may still have significant power in many respects. For example, despite the fact that manufacturing output has fallen by as much as 15 per cent since 1978, the decline in manufacturing employment relative to output has actually slowed down. The ratio of reductions in employment/reductions in output in manufacturing was 1.74 for the period 1973-8; for 1978-83, the ratio fell to 1.47. Second, despite the absolute fall in union membership, union density - at 50.6 per cent - is still higher than in the early 1970s (although the distribution of membership has changed significantly). Third, between 1978 and 1982 output per worker in manufacturing increased by only 6 per cent - that is, at an even lower annual rate than it had in the previous decade (it is true that in the first quarter of 1983, manufacturing productivity rose by 4 per cent relative to the last quarter of 1982 - but it is as yet unclear whether this is the start of a new trend). Fourth, while the cost of living increased by 71 per cent between 1978 and June 1983, earnings in manufacturing rose by 86 per cent. Fifth, the level of strike action - at least on some indicators - is no lower than in the 1960s and some years in the 1970s. (These trends are all the more striking when compared with the inter-war period, when union density collapsed rapidly, and real earnings fell.) Moreover, discussions with union activists and managers, along with findings from ongoing case studies, suggest that in many instances shop stewards continue to play an important and active role. The most publicised accounts of "macho" management relate primarily to the public sector - where top management are on short-term contracts and where major losses give the government a controlling role: but even in British Leyland not only are there signs that stewards are "regrouping", but also labour productivity - despite dramatic increases over the last few years - is currently no higher than it was in the late 1960s.

However, there are no systematic data available concerning the role and organisation of shop stewards. The latest survey - just published at the time of writing this paper - not only maintains the primarily procedural focus of most earlier surveys, but also was actually administered in 1980. Given the dearth of systematic evidence of a more recent kind, it was decided to undertake a survey, focussing upon the sorts of situation which had been the centre of debate over the last two decades - large plants in the private manufacturing sector. A grant from the Research Committee of the Department of Social and Administrative Studies at Oxford permitted a small postal survey to be undertaken: a sample, weighted according to size of labour force, was drawn from companies primarily or significantly involved in manufacturing which are listed in the Times Top 1,000 Companies. A sample of plants was then drawn from this sample of companies. In all, 312 plants were sent questionnaires, addressed to the personnel manager, in late June 1983. In all, about two thirds of the questionnaires were returned; rather less than a third of these declined to co-operate with the survey and in 4 per cent of cases manufacturing has ceased. The number of completed questionnaires was 133 - over 40 per cent of those initially sent out.

About one third of the completed questionnaires were from plants engaged in metal-handling and a further third from food, drink and tobacco and chemicals. Other industries with sizable representation included instrument and electrical engineering, textiles and clothing, and paper, printing and publishing. Only 8 per cent of the final sample employed less than 100 full-time manual workers, and over a third employed 500 or more. In all, the sample used in this analysis (124 plants) account for over 2 per cent of employment in manufacturing industry.

In these primarily large plants, both personnel management and trade unionism are well-established. 82 per cent have a personnel specialist at plant level, 73 per cent have a specialist industrial relations director at company level, and 68 per cent have a specialist personnel function above the level of the plant. In over half the plants the personnel function has increased in importance over the last five years, and in less than 5 per cent of plants has its role been reduced. Personnel and industrial relations considerations appear to play roughly as central a role in management decision-making as they did in 1978. In three-fifths of the plants manual union density is 100 per cent, and in a further quarter it is over 90 per cent. In only 3 per cent is there no union membership. In two-fifths of plants more than one manual union is recognised. Closed shop arrangements cover all the manual workers in just over half the plants; in a third of plants the closed shop covers no workers whatsoever. Check-off arrangements exist in just over 90 per cent of plants. In only one plant where a union is recognised are there no shop stewards, and in almost 90 per cent there are six or more shop stewards. The average size of manual steward constituencies was 28 members. These basic data suggest little change from 1978 (Brown, 1981). In the remainder of this paper, changes in managerial strategy are first discussed, and then the existing pattern of steward organisation is considered. (The findings of the survey are discussed more fully in Batstone, 1984.)

Management Strategy

Three fifths of the plants in the sample experienced a fall in output over the last few years; and it is likely that some of those which have increased output did so as a result of closures of other plants within their parent companies. Given this background, management strategies - or possibly more accurately, tactics - can be divided into four broad kinds - reductions in employment, attempts to change working practices, attempts to involve stewards and/or employees more fully, and endeavours to adopt a firmer approach towards employees and their unions or to reduce the role of the unions.

Only 12 per cent of plants have not reduced the size of their manual labour force over the last five years; 36 per cent reduced the workforce by 26-50 per cent, and 16 per cent by over half. In overall terms the sample of plants reduced its manual labour fore by slightly over a third over the last five years. Substantial reductions in manpower are more likely where output has fallen and the plant's financial performance is relatively low for the industry. Reductions in manpower are substantially more likely where trade unionism - despite the reduction in labour - continues to be a big or fairly big factor in obstructing more efficient working practices. While the range and amount of bargaining are less likely to have increased where there have been substantial reductions in labour, in only a quarter of cases have they fallen. However, large manpower reductions are associated with a tougher approach to the unions and with attempts actually to reduce the role of the unions.

Second, 85 per cent of plants have introduced changes in working practices over the last five years. In most cases, these have been associated with new capital investment. In about 70 per cent of cases these changes have led to major increased in productivity and efficiency. Such changes were the subject of negotiation in all but 10 per cent of plants. The inevitable constraints upon a postal questionnaire precluded a detailed investigation of these changes in working practices. In none of the plants where there were <u>no</u> changes in working practices are the unions seen as a big or fairly big factor in preventing more efficient working arrangements.

The third approach adopted by management has been an extension of the participation strategies initially introduced in the mid-70s. Respondents were asked whether management had adopted any major changes in their approach to shop stewards over the last five years. Two thirds say that they have done so: the single most common response (in half of the plants where changes had been introduced) is that stewards have become more fully involved and are now consulted more fully. In nearly nine out of ten plants formal consultation with union representatives existed prior to 1978; but the depth of consultation has increased considerably since then, largely - it would appear - as a means of impressing upon shop stewards the harsh realities of economic existence in an endeavour to achieve co-operation. Endeavours to increase steward involvement are slightly less common where no reductions in the manual labour force have occurred; they are associated with a greater role for stewards - the range and amount of bargaining has increased along with steward influence over members. Not surprisingly, where union involvement has been increased there is much less likelihood that management is trying to reduce the role of the unions. There is no tendency for increased union involvement strategies to be associated with trade unionism being a less obstructive force as far as working practices are concerned.

There have been attempts in a number of companies to increase employee involvement. Respondents were asked whether they had adopted any major changes in their approach to employees over the last five years. Two-thirds claim to have done so, and in three-quarters of these the changes include some form of increased involvement or consultation of workers. In about half of these plants union involvement has also been increased. A number of types of employee involvement were investigated more directly in the survey. Formal consultation with non-union representatives exists in just under a third of the plants, and is associated with lower levels of union density; its usage has not increased significantly over the last five years. Greater usage, however, is being made of other forms of worker involvement. Quality circles exist in roughly a fifth of plants and have generally been introduced since 1978. Briefing groups existed in nearly two-fifths of plants in 1978 and since then another fifth of plants have introduced them. Employee reports are distributed in three-quarters of plants and in a third of cases have been introduced in the last five years. Almost a fifth of plants have autonomous work groups; slightly more than half of these were introduced prior to 1978.

In all, one or more of the most direct strategies of employee involvement - quality circles, briefing groups or autonomous work groups - exist in four-fifths of the plants employing 200 or more workers, and in only half the smaller plants. Briefing groups and employee reports are rather more common where "bureaucratic" steward bodies exist, while quality circles are more common in less "bureaucratic" situations. There is no significant tendency for direct employee involvement strategies (i.e. briefing groups/quality circles/autonomous work groups) to be associated with variations in union strength and control.

In 20 per cent of plants moves have been made over the last five years to increase employee involvement without making any similar moves as far as shop stewards are concerned. These tend to be plants which do not bargain at establishment level. Such strategies, respondents claim, have affected the role and influence of shop stewards: indeed, increases in employee involvement alone are closely associated with attempts of a more direct nature to reduce the role of stewards. There has been a "net reduction" (i.e. percentage of cases where X had been reduced, less percentage of cases where X had been increased) of steward influence over members of 33 per cent in these plants compared with a figure of only 6 per cent in other plants. Similar, though less sizeable, reductions in senior steward influence are also found where recent attempts to increase employee involvement have been made. Moreover, a net reduction in the amount of bargaining of 20 per cent has occurred in plants increasing employee involvement, compared to a small net increase in other plants. Despite these moves, however, increased employee involvement is not significantly associated with lower levels of steward influence over members or over the organisation of work. Indeed, despite the fact that productivity growth relative to output tends to be higher in these "new employee involvement" plants, they are still more likely, according to respondents, to be prevented from adopting more efficient working practices due to union power: this is so in three quarters of these plants compared with less than half the others.

Fourth, some managements are adopting a firmer line with stewards and workers (in about a fifth of plants) and, in 15 per cent of cases (primarily where a firmer line is also adopted) attempts are being made to reduce the role of the unions. In only one case, however, has shop steward organisation - of a weak kind and in a relatively small plant - actually been removed. Such strategies are rather more common in industries which are suffering particularly severely from the depression and in plants whose financial performance (compared with the industry average) has been relatively poor over the last few years. They are also associated with a reduction in the influence of personnel specialists (although such specialists are no less likely to exist). Attempts to reduce the role of the unions - as distinct from simply adopting a tougher line towards stewards and workers - tend to be found where steward organisation is less bureaucratic, where bargaining occurs above the level of the plant and where unions are more of an obstruction to efficient working practices.

The adoption of a tougher approach by management appears to have a number of effects. Steward influence over members has experienced a "net reduction" of about half where management has pursued this strategy compared with a "net reduction" of only 3 per cent elsewhere (senior steward influence has decreased less dramatically). Along with this, stewards have experienced a "net reduction" in the range and amount of bargaining and in their level of influence over work organisation where management has adopted a firmer approach, compared with a "net increase" in other plants. While in most plants management have tried to give formal, written agreements a more central role, in cases where a tough line has been adopted there is a tendency for agreements to have less significance - presumably because management is seeking to reduce the extent of joint regulation. However, despite these "successes", managements who have adopted "the firm approach" are still more likely to be faced with union obstruction of more efficient working methods and to have experienced some form of industrial action over the last two years.

It should be remembered, however, that concerted attempts to challenge the unions are occurring in only a minority of plants. This is not, of course, to suggest that managements are not adopting courses of action which are unpalatable, notably redundancies, elsewhere (although, particularly where steward organisation is strong, redundancies have been primarily voluntary rather than enforced). Reductions in manpower, and flexible working practices - predominantly negotiated - are the main strategies adopted. In addition, many plants are trying to increase the involvement of both stewards and workers, relying upon economic realities to win co-operation. Moves to increase employee participation, which in various forms are very widespread, do of course constitute a challenge to the traditional monopoly of stewards over worker-management communication in many instances. But - at least to date - it appears to have a relatively limited impact even where there are no corresponding moves to increase the involvement of stewards. (There were very few cases where management had changed the systems of pay determination over the last five years.)

Organisation and Operation of Shop Stewards in 1983

I now consider the extent to which steward bureaucracy survives and the role it plays. Since senior stewards exist in virtually all plants where there are stewards, only two aspects of bureaucracy will usually be considered below - the existence of full-time stewards and of regular steward meetings.

Regular steward meetings are held in just over half of the plants where shop stewards exist; in just under a third of plants where there are two or more manual unions, regular joint shop steward meetings are held. Two fifths of plants have full-time shop stewards. A third of the plants have fully bureaucratic steward organisations and a further quarter have two bureaucratic features. Only one of the 34 plants with fewer than 200 manual workers is fully bureaucratic while, at the other extreme, two-thirds of those with 500 or more manual workers have all three features of steward bureaucracy. 15 per cent of the smallest plants have no bureaucratic features as do 5 per cent of the largest plants. Once size is controlled, there appears to be little significant variation between industries. Full-time stewards exist in 59 per cent of cases where there is corporate bargaining above plant level, in 38 per cent of cases where there is plant bargaining, and in only 23 per cent of others. In just under two-thirds of plants where there is intra-company bargaining (at or above plant level) there are regular steward meetings, while in only 29 per cent of other plants did stewards meet regularly.

David Deaton has adjusted the 1978 Industrial Relations Research Unit survey to make it comparable with the 1983 survey. These data indicate that the degree of steward bureaucracy in large subsidiary plants has declined only marginally over the last five years, despite massive reductions in employment. This is in stark contrast to experience in the inter-war years. But the thesis of steward incorporation would explain this by arguing either that bureaucratic steward organisations are dominated by a steward elite which promotes management rather than worker interests, or that these organisations have become empty shells. In seeking to assess this thesis, it is necessary, first, to assess the distribution of power within shop steward organisations and, second, to consider the amount of control over labour relations which they exert and the uses to which they put such control.

a) The Distribution of Power

From the survey data it is possible to investigate four aspects of the distribution of power within shop steward organisations. The first concerns that classic indicator of oligarchy, namely, tenure. The other three relate to different areas of control - the distribution of power within the shop steward body as such: the degree of influence stewards exert over ordinary union members/workers and whether senior stewards tend to exert more or less influence than "ordinary" stewards; and the relative influence of stewards and senior stewards over the organisation of work and members' earnings.

If steward bureaucracy encourages oligarchy, then one would expect that steward tenure, and particularly the tenure of senior stewards, would be greater where bureaucratic characteristics exist. In fact, steward bureaucracy appears to have rather complex effects on steward tenure. Regular steward meetings tend, overall, to reduce tenure; on the other hand, the existence of full-time stewards reduces the tenure of senior stewards and increases that of other stewards, but it is nevertheless associated with a strong tendency for senior stewards to enjoy longer tenure tan other stewards.

The second aspect of oligarchy distinguished above concerns the distribution of power within shop steward organisations. In order to investigate this question respondents were asked whether they would describe the way in which shop stewards operated in their plant as either "senior stewards exert tight control over other stewards", or "stewards collectively formulate an overall policy binding on them all", or "individual stewards and work groups tend to go their own way". Individual steward autonomy occurs primarily where no features of steward bureaucracy exist (50 per cent of cases), and where there are only regular meetings of stewards (31 per cent of cases). Collective control exists in the majority of plants except where no bureaucratic features are present. Control by senior stewards exists most frequently where all bureaucratic characteristics are found (27 per cent of cases), but the variations in this respect are not very large. In sum, while bureaucracy may reduce steward sectionalism, the extent to which it increases steward oligarchy - according to the survey - is rather limited.

The third dimension relevant to the question of steward oligarchy concerns relationships between stewards and members. Shop stewards are seen to have a substantial amount of influence over members in the vast majority of cases - it tends to be least where no bureaucratic features exist. Hence in only 7 per cent of plants with no steward bureaucracy are stewards said to have a "great deal" of influence, while this was so in 41 per cent of plants with a full steward bureaucracy. Regular steward meetings, and full time stewards even more so, are associated with a "great deal" of steward influence over members. In short, bureaucracy is associated with a tendency towards greater steward influence over members.

Respondents were also asked whether senior stewards tend to have more influence over members than other stewards. This is reported to be the case in more than half the plants. However there is no significant tendency for this to vary according to the degree of steward bureaucracy. In short, while bureacracy tends to increase steward influence over members, it does not tend to increase senior steward influence - hence, in this respect, it is not oligarchical. (It is worth noting here that stewards are able to hold meetings of members in 90 per cent of the plants, and in 70 per cent can do so during working time. Moreover, stewards can leave their jobs in 97 per cent of plants; in only just over a quarter of cases is permission for stewards to leave their jobs ever refused. In other words, in the great majority of plants there appear to be few problems placed in the face of steward-member communication; this is particularly so given the high levels of steward density).

The fourth aspect of oligarchy which can be investigated through the survey concerns the relative influence of stewards and senior stewards over the organisation of work and their members' earnings. The actual levels of influence are considered in the next section: here the focus is upon the relative influence of the two types of stewards. As far as work organisation is concerned, the "net" proportion of plants in which senior stewards are deemed to have greater influence than other stewards is 21 per cent. There is no significant tendency for senior stewards to have greater relative influence where there are full-timers. Where regular steward meetings are held, however, the "net" figure for greater senior steward influence rises to 29 per cent, while for the remainder it is 10 per cent.

Senior stewards are more likely to have greater relative influence as far as members' earnings are concerned - the "net" figure is 34 per cent. While the existence of full-time stewards only marginally increases the relative influence of senior stewards, the holding of regular steward meetings has a greater impact. The "net" figure rises to 44 per cent where such meetings are held, while for the remaining plants the figure is only 18 per cent.

The survey evidence, then, indicates that the existence of bureaucratic characteristics does not lead to any simple and progressive trend towards oligarchy. In some cases, for example some aspects of tenure, it actually reduces it, although overall, there is a weak tendency for bureaucracy to increase the relative control of senior stewards. Moreover, as the next section indicates, the relatively greater power of senior stewards in bureaucratic situations is associated not with less, but more, influence on the part of other stewards. In brief, the findings of the survey give only a small measure of support to the oligarchy thesis.

The Level and Use of Steward Control Over the Labour Process

Now I consider the second theme of the steward incorporation thesis, namely, the extent to which stewards influence the wage-effort bargain and the uses to which they put that influence. This thesis argues that bureaucratic steward organisations have only limited influence over the wages and conditions of work of their members; that any such influence is concentrated at the level of senior stewards rather than normal stewards; and that steward activity is both shaped and constrained by formal, written agreements. These features, it is argued, derive from a sympathy with management as distinct from worker interests, so that bureaucratic steward organisations rarely, if ever, resort to industrial action or obstruct management attempts to introduce efficient working practices. This section therefore considers, first, the extent of steward and senior steward influence over earnings and work; second, the role of formal, written agreements; third, the extent to which stewards and senior stewards adopt a managerialist perspective; fourth, the level of obstruction of management ends. Finally, the extent to which stewards have developed forms of organisation which might permit them to influence corporate decision-making is briefly considered.

In the survey respondents were asked to rate the influence of both stewards and senior stewards in determining work organisation and manual workers' earnings. In a quarter of plants negotiations by senior stewards are rated as "very important" in determining the organisation of work, and in a further two-fifths it is seen as "fairly important". While the existence of regular steward meetings marginally reduces the rating of senior steward influence, the existence of full-timers tends to increase it. Hence, in 43 per cent of plants where there are full-time stewards negotiation by senior stewards is seen to be "very important" in determining the organisation of work, while this is so in only 12 per cent of other plants.

However, the existence of full-time stewards is also associated with a more important role for "other" stewards in determining work organisation. In two-thirds of the plants with full-timers, negotiation by stewards is seen to be "very" or "fairly" important in affecting work organisation, while this is so in only two-fifths of other plants. There is no statistically significant relationship with regular steward meetings.

Negotiation by senior stewards is seen to affect the earnings of manual workers a "great deal" or a "fair amount" in 56 per cent of plants. The impact of steward bureaucracy is less great than in the case of work organisation. In 63 per cent of plants with regular steward meetings and/or full-timers, senior stewards are seen to have a substantial impact on members' earnings, while this is so in 43 per cent of other plants. As far as the significance of "other" stewards for workers' earnings is concerned, the existence of bureaucratic features has virtually no effect.

Another means of looking at the impact of bureaucracy on steward control is to consider the issues which stewards negotiate. Respondents were asked whether stewards negotiate six issues - pace of work, manning levels, transfers between jobs, contracting work out, redundancy, and discipline and dismissal (this was a particularly "tough" question - it did not ask whether stewards "ever personally settled" issues nor about negotiations "over issues relating to" these topics). The negotiation of the last two issues is common in most plants and is not significantly related to steward bureaucracy. However, negotiation over the other four issues, relating to the organisation of work, is rather less common. For example, the pace of work and manning levels are subjects of steward negotiation in almost twice as many plants with full-time stewards as in those which do not. Stewards bureaucracies - at least those with full-time stewards - are, then, more, rather than less, likely to be involved in the organisation of work.

The incorporation thesis, however, would suggest that even if bureaucratic steward organisations are involved in the organisation of work, they will be tightly bound by formal, written agreements which preclude negotiation and promote management interests. Certainly formal, written agreements tend to be more common where full-time stewards exist; they cover three or more issues in three-fifths of such plants but do so in only two-fifths of other plants. But when we look at each of the six issues separately, plants with full-timers are not more likely to have formal agreements over redundancy or discipline and dismissal, but are more likely to have them for work organisation issues. But in these cases they are associated with bargaining rather than acting as a substitute for it: hence, negotiation on these issues occurs in over two thirds of the cases where they are covered by formal, written agreements, but in only 13 per cent of other cases.

The role of formal, written agreements was also investigated more directly by asking which set of options best described their role "in relation to negotiations by shop stewards". Only 16 per cent of respondents say agreements play a negligible role or merely provide a broad framework for negotiations; 36 per cent say they "provide a set of rules over which a great deal of negotiation occurs"; and 48 per cent that they "limit and shape" or "effectively preclude negotiations". Formal agreements play a greater role in more bureaucratic situations, although where both full-timers and regular meetings exist they are less likely to constrain negotiations than where only one bureaucratic characteristic is found. Moreover, it should also be noted that the existence of more detailed agreements is associated with an increase, rather than a reduction, in steward control over work organisation and earnings. For example, in nearly two thirds of plants where agreements constrain negotiations, senior stewards have a significant degree of control over the organisation of work while this is so in only two fifths of plants where agreements play a negligible role or merely constitute a broad framework; a similar relationship is found with "other" stewards' influence over work organisation and earnings.

However, it is possible that stewards use their control to promote management as against worker interests. This brings us to the third question posed at the beginning of this section: namely, the perspectives which stewards tend to adopt. The survey used a question which had been asked in the 1966 Workplace Industrial Relations Survey - whether stewards were inclined to take management's or workers' views or took a 50/50 approach. In 1966 about three-fifths of managers said stewards took the workers' view and the vast majority of the remainder said their view was "50/50". The figures from the 1983 survey are virtually identical. Moreover there is no significant tendency for stewards in bureaucratic organisations to show greater sympathy with management views. Respondents were also asked whether senior stewards are more or less sympathetic than other stewards to management's views. As might be expected, senior stewards are seen as more sympathetic in just over half the plants, and less sympathetic in 10 per cent of cases. Senior stewards in more bureaucratic organisations tend to be only slightly more sympathetic. Support for the "bureaucracy = incorporation" thesis, then, is rather weak as far as stewards views are concerned.

The basic issue, however, for the steward incorporation thesis is the extent to which stewards (and workers) actually use their power to obstruct management. Two sets of questions are relevant here: the first concerns the experience of industrial action over the last two years, the second the extent to which trade unionism prevents the adoption of more efficient working practices.

61 per cent of the plants have experienced some form of industrial action over the last two years, and 40 per cent strike action. The use of sanctions by workers is considerably more likely where steward bureaucracy exists. For example, 60 per cent of plants with full-time stewards have experienced one or more strikes, compared with only 28 per cent of the remaining plants. Similarly, four-fifths of plants with full-timers have had some form of industrial action, while less than half the other plants had.

In order to investigate the extent to which trade unions constrained managerial freedom, a question from the 1966 survey was used. Managers were asked whether there were any ways in which the organisation and arrangement of work in the plant could be improved if they were free to arrange their labour force as they wished. Where managers felt this was possible, they were asked what percentage of working hours could be saved, and the significance of trade unionism as a factor "preventing you from arranging your labour as you would wish". In 1966 about half the managers (and 64 per cent of those in large plants) believed that working practices could be made more efficient; just under a third of those who felt able to specify potential savings put the figure at 11 per cent or more; and just under half thought that trade unionism was a big or fairly big factor in preventing more efficient practices (Government Social Survey, 1968:90).

In 1983, 62 per cent of managers feel that working practices could be improved, 44 per cent claiming potential savings of 11 per cent or more. 73 per cent of managers who see the possibility of greater efficiency say that trade unionism is a big or fairly big factor in preventing changes in working practices. Steward bureaucracy, in the form of full-time stewards, is associated with a greater probability both that managers believe labour savings are possible (68 per cent compared with 54 per cent), and that trade unionism is a significant factor in preventing the adoption of more efficient labour practices (60 per cent compared with 34 per cent). Estimated labour savings also tend to be greater where elements of steward bureaucracy exist.

Finally, there has been considerable interest over recent years in the ability of shop stewards to influence corporate policy above the level of the plant. Combine committees are often seen as an important means by which stewards might be able to do this. In fact, combines typically constitute little more than opportunities for exchanging information (primarily about wages and conditions) and have an unfortunate tendency to collapse if anything more is asked of them, particularly where pay bargaining is at plant level. However, accepting the assumption that combines provide stewards with a greater ability to influence corporate policies, their existence was investigated in the survey. In only 10 per cent of plants do stewards meet regularly with their counterparts from other plants in the company, although they do so occasionally in a further third of cases (this suggests some decline since 1978). Participation in a combine is closely associated with steward bureaucracy: stewards are members of a combine in half the plants where one or more features of bureaucracy exist, but in only a quarter of other plants.

In this section, the focus has been upon the degree of control which stewards exercise and the purposes to which they put it. It was found that in more bureaucratic organisations both senior and other stewards (except in the case of earnings) exert greater control, and that a larger number of work organisation issues are negotiated. Bureaucratic stewards are more likely to be associated with more detailed agreements, but these are found generally to be associated with a greater range of negotiation and steward control. There is no tendency for stewards in bureaucratic organisations to demonstrate greater sympathy with management, although there is a slight tendency for senior stewards in bureaucratic bodies to be more sympathetic to management than other stewards. Finally, both industrial action and the "obstruction" of more efficient working practices is more common in bureaucratic bodies, which are also more likely to be involved in a combine committee.

In this paper I have argued that the adverse conditions for "corporatism" which had always existed were in certain respects intensified during the 1970s. For while the reform of workplace industrial relations led to an extension of shop steward organisation and some increase in steward bureaucracy, it did not lead to the general incorporation of shop stewards. This fact had both political and economic implications, which played a part in the move to the current situation of massive unemployment and a hostile political environment for trade unionism.

In this new context, managements have adopted a variety of techniques, notably reductions of manpower, changes in working practices and attempts to increase union involvement and worker involvement: in a minority of cases concerted attempts have been made to remove, or reduce, the role of shop stewards. Clearly these strategies have met with some success and it would be absurd to argue that shop stewards have not experienced a reduction in their power and influence. But the survey evidence indicates that, even in the face of these new (or newly emphasised) management strategies, there is little evidence that bureaucratic steward organisations have become especially oligarchic or incorporated. It is, of course, quite possible that in the future the situation may change, but - as yet - it is far too early to talk of the widespread development of a form of enterprise unionism similar to that in large private companies in Japan.

It should be stressed that the discussion has focussed upon manual workers in predominantly large manufacturing plants which are owned by large companies - the focus of industrial relations analysis for the last twenty years. There is also reason to treat the findings with some care: they are based on a relatively small sample (although it covers a not insignificant proportion of the manufacturing labour force), and the survey was inevitably rather general and solicited information from management (but these features are far from unique to this study). But given the size of labour force reductions in the sample, it can scarcely be argued that the stewards investigated have had a uniquely easy time over the last five years.

With these cautions and limitations in mind, the findings suggest that in the "classic Donovan" situations studied trade unionism has demonstrated the strengths and weaknesses which it has shown since the Second World War. What appears to have happened is that during the decade after Donovan, shop stewards increased their power and were associated with a further worsening of the competitiveness of British industry. Over the last five years management has recouped some control but - so far - it seems plausible to suggest that in many cases stewards still play a significant role and exercise a not inconsiderable influence over traditional industrial relations issues.

REFERENCES

Batstone, E. 1984: Working Order: Workplace Industrial Relations in Britain Over Two Decades, Oxford.

Batstone, E. 1981: "Industrial Democracy in Britain: The Rationale and Fruits of Reformism", in H. Diefenbacher and H.G. Nutzinger, eds., Mitbestimmung: Problems und Perspectiven der empirischen Forschung, Frankfurt.

Batstone, E., A. Ferner, and M. Terry: Consent and Efficiency, Oxford.

Brown, W., (ed.) 1981: The Changing Contours of

British Industrial Relations, Oxford.
Currie, R. 1979: Industrial Politics, Oxford.
England, J. 1981: "Shop Stewards in Transport House: a Comment on the Incorporation of the Rank-and-File", Industrial Relations, xii (5).
Government Social Survey, 1968: Workplace Industrial Relations, London.
Hart, H. and C.V. Otter 1976: "The Determination of Wage Structures in Manufacturing Industry", in R. Scase, ed., Readings in the Swedish Class Structure, Oxford.
Hyman, R. 1979: "The Politics of Workplace Trade Unionism", Capital and Class, 8.
Kahn-Freund, O. 1979: Labour Relations: Heritage and Adjustment, Oxford.
Korpi, W. 1978: The Working Class in Welfare Capitalism, London.
Pizzorno, A. 1978: "Political Exchange and Collective Identity in Industrial Conflict", in C. Crouch and A. Pizzorno, eds., The Resurgence of Class Conflict in Western Europe since 1968, London.
Sabel, C. 1983: "The Internal Politics of Trade Unions", in S. Berger, ed., Organising Interests in Western Europe, Cambridge.

Chapter Eight

SHOP STEWARDS AND MANAGEMENT: COLLECTIVE BARGAINING AS CO-OPERATION

Michael Terry

Much ink has been spilt in debate over alleged recent changes in shopfloor politics, attitudes, and practices. Is there a "new realism" of co-operation rather than conflict between workers and employers, based on a shared perception of the bleak consequences of failure? Are managements exploiting the bargaining advantage afforded by recession and unemployment to "roll back" what they see as the more obstructive and damaging aspects of "traditional" shopfloor trade union organisation? What is the significance of the recent resurgence of joint consultative structures in industry and of other techniques for disseminating information from employers to workers?

Part of the problem in seeking sensible answers to such questions lies in finding an adequate characterisation of the relationships which existed prior to these recent alleged changes. Discussions on this topic have suffered in particular from a simplistic treatment of the union-management relations involved in "traditional" processes of collective bargaining. This paper attempts a sketch of collective bargaining against which recent developments may be more adequately evaluated.

Collective Bargaining: Consultation and Negotiation

Collective bargaining has often been held to involve opposing interests; it is seen as an institutional device for accommodating antagonistic, conflictual relationships. On this basis it has often been contrasted with more co-operative, "unitary" organisations such as works councils found in some other European countries. This paper proceeds from the assertion that while there are certainly points of contrast between the two systems, the characterisation of collective bargaining as essentially antagonistic is over-simple and misleading.

Within the field of collective bargaining a distinction is often made between consultation and negotiation and this distinction is again becoming important. Consultation supposedly differs from negotiation because it concedes no right - de jure or de facto - to trade unions to be involved in taking decisions. Conversely negotiation formally allows unions to take a view that may differ from that of management, to press it, and ultimately to register disagreement, with the possibility of taking action to press their case further. Where a <u>status quo</u> arrangement exists, management may be effectively prevented - if only temporarily - from implementing decisions. Consultation, on the other hand, is characterised as "co-operative" in nature, involving workers and managers in a process of mutual problem-solving and information handling. It explicitly disallows trade unions the right to participate in decision-making, consisting instead of informing unions of managerial decisions and seeking responses.

Thus within the British system the set of formal bargaining relationships between managers and unions covers both allegedly "co-operative" and "antagonistic" structures. For several years the conventional wisdom of British industrial relations has emphasised the process of negotiation rather than consultation, the former being seen as the fount of the "joint regulation" of work relations that allegedly characterises much of the unionised British industry. This paper seeks to challenge this conventional wisdom in so far as it obscures the essentially collaborative, co-operative, set of relationships on which much of daily collective bargaining is based - whatever it may be called formally. This is not to deny the existence of antagonisms and their expression, on occasion, through bargaining; merely that the norm of industrial "peace" implies co-operation of a kind, and this is reflected in management/union relationships.

<u>Collective Bargaining: The Basis of Consensus</u>

Two distinct aspects of collective bargaining may be considered to act as practical constraints on the process of negotiations. The first is a limitation on the scope of collective bargaining; and the second involves the nature of the procedures and understandings on which the process is based, both formally and informally.

Advocates of collective bargaining as the appropriate means for extending worker control over managerial behaviour freely concede its relative lack of success hitherto in extending to strategic areas of managerial decision-taking. For them the importance of the device is summed up in the well-known quote from Hugh Scanlon that "the best form of industrial democracy is the extension of collective bargaining to which we know no limit". The crucial element is the last phrase. The nagging inability or reluctance of unions to expand the scope of collective bargaining does raise questions about its real potential. But the hesitation of unions is more than matched by the opposition of managers. When Lucas Aerospace shop stewards attempted to raise strategic management questions for negotiation through their "Corporate Plan" they were met by a blank refusal from management. Other unions have found initiatives frustrated by management's assertion that certain topics are not open to negotiation, an assertion that is echoed in the statutory insistence that trade unions are only entitled to information which can be held to be necessary for the purposes of collective bargaining. This limitation builds restrictive circularity into the relationship between information and the scope of bargaining. Management practice has been to seek to restrict the scope of collective bargaining, and to deny the "Donovan" philosophy that all issues are potentially open to negotiation (see Ogden, 1981).

Fox's suggestion (Fox, 1973) that employers may well concede the principle of joint agreement "on the margins" of their concerns, but strongly resist union attempts to invade central strategic decisions remains a reasonable hypothesis. This has contributed to a situation in which there is widespread de facto agreement between unions and employers as to what constitutes the proper scope of collective bargaining. Thus it has been suggested that formal definition of the issues which are negotiable acts to define other issues as beyond the competence of bargaining, thereby making it more difficult for unions to mount the necessary ideological and practical arguments to extend bargaining scope (see e.g., Sisson, forthcoming). Outside the relatively narrow focus of collective bargaining, unilateral regulation (usually by management, sometimes by workers) exists (see e.g., Storey, 1983, p. 99). And areas of unilateral managerial action may often, even now, include hiring and firing (subject to statute), mobility of labour, and other issues of daily importance to workers. Thus unions "co-operate" with management in so far as they accept "management's right to manage" to a considerable extent.

But of at least equal importance to the restrictions on the scope of collective bargaining are considerations of the nature of collective bargaining itself. Hyman (1971) contrasted two kinds of union strategy to influence members' conditions of work: bilateral and unilateral job regulation. Of the former he argued that "the union which adopts this strategy must seek an amicable relationship with employers through a readiness to accept compromise settlement of its demands, caution in exploiting positions of temporary bargaining strength, and strict adherence to agreements reached". And this is not all. Hyman went on to argue that "responsible trade unionism involves the acceptance of an obligation to control the rank and file ... it requires the union to exercise quasi-managerial discipline against members who rebel against an oppressive work relationship" (Hyman, 1971, p. 225). While this analysis leaves scope for variation in the extent of compromise and co-operation - allowing at least that an increase in worker bargaining strength may at least give them a "more acceptable" compromise - it nevertheless makes a powerful critique of the assertion that collective bargaining continues to reflect and assert the antagonistic relationships of which the existence of trade unions is an expression.

Collective Bargaining as Co-Operation

Perhaps the simplest starting point is to present collective bargaining as a process which, from the point of view of management, presents both advantages and drawbacks. The attitude of management to collective bargaining is not static, it varies according to such features as labour and product market pressure, pressures for innovation, and the extent to which there is political support for trade unions and collective bargaining activities. Thus, although as Ogden demonstrates, many managers never accepted the Donovan Commission's arguments for the extension of joint control of decisions (Ogden, 1981), there is little question that the late 1960s and 70s saw an extension of collective organisation and trade union recognition, especially at plant and company levels. This was facilitated, if not actively encouraged, by managers who perceived at least some advantages in the operation of formal collective bargaining arrangements (Terry, 1983; Marchington & Armstrong, 1983). More recent events suggest that some companies may be moving away from that strategy. However, before examining recent developments, it is worth setting out at least some of the reasons for and consequences of changing patterns of managerial acceptance of the relative merits of collective bargaining.

The immediate post-war period witnessed the development of shopfloor trade union organisation in some parts of manufacturing industry, especially engineering. Certainly not encouraged, and often bitterly opposed, by employers, this unionism sought to introduce a degree of stability into the fluctuating and uncertain world of piecework, changing technologies, and rising but varying demand. Although the scope of bargaining varied, its central concern was wages. In a period during which companies were generally able to pass on wage increases into prices, this seemed at least tolerable (although in companies such as Ford, with centralised wages agreements, stewards were more strongly resisted than elsewhere). And in exchange, employers and, perhaps more importantly managers, began to appreciate the advantages that could accrue from the orderliness and immediacy of plant-level industrial relations (see, e.g., Turner et al., 1967, p. 214). The advocacy of collective bargaining as a device helping both management and unions in problem resolution found its clearest expression in the 1968 Donovan Report.

Experience in the 1970s suggested that this mutually beneficial exchange could be deepened. Both Brown (1973) and Batstone et al (1977) demonstrated that, by and large, the better integrated the shop stewards organisation, and the tighter the control over sectional activity, the more successful was the stewards' pursuit of wage claims. The resulting pressures towards more centralised shop stewards organisations has been thoroughly documented. One associated consequence was an increase in shop steward "leadership", with stewards increasingly characterised as the initiators of activity rather than being the mouthpiece of a group of which they were a member. Various writers have described this as a move from "direct", "primitive" or "participatory" democracy to one of "representative" democracy controlled through elections. One key mediating influence in this process has been termed the "bargaining relationship", and this too operates in a manner that could be called "co-operative".

Without implying shared substantive aims, the existence of a "strong bargaining relationship" between particular managers and shop stewards (usually senior ones) means that they are "on the same wavelength", sharing understandings and definitions. In this way they make the negotiations that take place "sometimes meaningless (to the outsider) for they simply consist of shorthand cues. But bargaining therefore becomes more effective and is based on 'private rules'" (Batstone et al., 1977). The co-operation between managers and stewards that this relationship fosters is designed to try to strengthen the position of each of the parties. In other words, senior stewards may try and avoid embarrassment to the managers by working to avoid overt signs of industrial conflict; managers may work to strengthen certain stewards with respect to others by the judicious granting of concessions. This is the bargaining of shared rules and understandings, of the philosophy of "swings and roundabouts" - "you win some; you lose some". Stability and continuity are the keynotes, and sudden or dramatic change is difficult to accommodate.

Batstone and his colleagues suggest that such bargaining relationships are only to be found where "there is a broad balance of power between the two persons involved". More recent evidence suggests, however, that similar kinds of relationships appear to exist in situations where by most objective criteria the ability of unions to inflict damaging economic or political damage on the employer is lower than in the case studied by Batstone et al. In some cases there is evidence that management themselves were the prime movers in encouraging or fostering a system of shop floor representation in which the pivotal relationship was between personnel management and one (or very few) senior stewards/convenors with considerable personal authority and control over their membership, much of it derived from their position as accorded by management. The stability and "order" associated with local trade union organisation and powerful central leadership were pursued by management during this period as useful precursors to the negotiated introduction of a range of changes. In these cases, as opposed to those with traditions of strong shopfloor organisation, negotiation with senior shop stewards was seen as likely to facilitate rather than hinder the introduction of change, at least in the short-term. That these objectives continue may be seen in, for example, the support given by management to the establishment of closed shops (Brown, ed., 1981) and to the opposition of such bodies as the Institute of Personnel Management and the Engineering Employers Federation to the Employment Act 1982 proposals to weaken the closed shop. During the 1970s the bogey for managers was "fragmentation" in trade union organisation, and they were prepared to concede considerable formal rights to trade unions in an effort to avoid or reduce it.

There are good reasons - not necessarily the same - why management and unions should have found themselves in agreement on the desirability of more formal, secure, and centralised trade union structures. For management the major advantage was the chance of securing consent from the steward organisation to their proposals, and then attempting to secure the co-operation of the steward organisation itself for the implementation in the event of dissidence - the so-called "policing of agreements".

Grievance and other procedures also operate to encourage change within union organisation of a kind that might have some advantages for management. Being devices predicated upon the existence of hierarchy within the organisations party to them, they can be utilised to encourage the development of hierarchical levels within unions. These can in turn operate to reduce the likelihood of "sectional" activity. The use of procedures creates precedents in a manner analogous to the courts. Thus, for example, certain types of "excuse" for absence from work, or lateness are established as "acceptable" in a particular company, others are not. This is one of the many ways in which the steward body comes to act as a "filter device" for grievances - using past practice to justify their refusal to put certain issues into procedure however strongly the worker or workers concerned might feel. Such novel claims that might be taken up would be judged in terms of their chances of success, or generally more in strategic terms, knowing how procedures operate, than through considerations of "inherent merit".

Shop Stewards and Management

As noted earlier, few unions in practice expressed or pursued policies and objectives that conflicted with "management's right to manage". Through the operation of, for example, disciplinary and dismissal procedures, unions did not contest management's fundamental right to hire and fire; they merely sought that the practice should be applied in a consistent and "fair" manner. To contest the basic right would have required the deployment not only of considerable union power but also of a range of political arguments of a kind that unions in this country have rarely expressed. For this reason bargaining in this country has largely taken place over issues that do not involve a fundamental clash of interests. Wages are the best example.

Except in periods of crisis, wages are rarely "all-or-nothing" issues. They occupy a continuum, along which "victory" or "defeat" are judged according to which side of the mid-point between claim and offer the final outcome lies. It is one of several issues that facilitate "face-saving" and compromise for both parties. Other areas of managerial decision, however, may involve "all-or-nothing" decisions. Should a worker be sacked? Should a factory be closed? Should the company open a factory in South Korea? and so on. Here the conventions of collective bargaining - compromise, and "win some, lose some" - are more difficult to apply. On these issues collective bargaining has made few inroads, and attempts that have been made have frequently met with managerial intransigence and, sometimes, official union disapproval. They are, as we shall see below, however, increasingly issues for "consultation".

Shop Stewards and Job Controls: A Challenge to Managerial Prerogative?

Worker- or union- enforced influence or control over the processes of production have frequently been cited as the example *par excellence* of shopfloor antagonism to managerial controls and the assertion of an alternative approach. "Restrictive practices" has been for over two decades a term of abuse and disapprobation, perceived as the antithesis of efficient working and "responsible" trade unionism. Certainly many trade unions have an ambivalent approach to such controls. The acceptance of the notion of a "fair day's work for a fair day's pay" implies that work restrictions can only be justified, if at all, on the grounds that the wages are exploitative. Where unions have negotiated the agreements, and encouraged their ratification, they have problems in accepting the agreed rates as "unfair". Nevertheless, they are also aware that job controls exist, persist, and form an important element in shopfloor behaviour. The ambivalence was perhaps best recognised by the Donovan Commission, albeit unintentionally, in its implied contrast between restrictive practices and other aspects of shopfloor bargaining. The latter, largely conducted by shop stewards, was perceived as the basis of beneficial future developments. The former, by contrast, was only grudgingly seen as in the interests of workers, "narrow and shortsighted though these interests may sometimes be", and shop stewards were, extraordinarily, not mentioned once in the discussion on restrictive labour practices. The implied argument that stewards were somehow outside this process was - and is - clearly untenable. The most likely explanation would appear to be that the Donovan Commission, keen to propound the development of mutually beneficial collective bargaining based around shop steward organisation, was reluctant to include within its ambit issues which either operated against the long-term interests of both workers and employers or which, more worryingly, might suggest the possibility of more deep and irreconcilable conflicts of interests than was implied by bargaining over wages.

More recent analyses tend to assign a more problematic status to job controls, implying at the least that not all job controls always work against the interests of employers. At one obvious extreme, the newspaper employers of Fleet Street were able to tolerate for many years a system in which virtually total union control of hiring, firing, manning and other practices obtained. While it is possible to see this as an anomalous case, it is less easy to dismiss, for example, the conclusions of Hyman and Elger (1981). Examining the development of job controls (and other manifestations of shopfloor worker activity) they argue that many previous analyses "fail to grasp the extent to which shopfloor organisation has followed the contours of British capitalist development, building quite limited and sectional gains without substantially circumscribing employers' room for manoeuvre, sometimes indeed furnishing important elements of self-discipline and collaboration with management"

(Hyman & Elger, 1981, p. 144). In other words, "workers control" may, through the need for trade union discipline and order to maintain its effectiveness, come to constitute for management a less costly means of securing the attainment of managerial goals than a strategy which seeks to undermine "workers control" and replace it with managerial controls. None of this is to argue that in pursuing such strategies workers and their representatives are self-consciously adopting collaborative approaches. Rather it implies that such approaches are in part a logical consequence of the adoption and pursual by unions of a policy of "joint regulation", even when the formal terms of such joint regulation do not cover the substantive areas allegedly regulated by "job control".

Equally obviously, however, not all "job controls" can be treated as unvaryingly co-operative in their consequences. Recent work has concentrated attention on the way in which changes in the material position of firms can transform the practical significance of "job controls" (see e.g. Hyman & Elger, 1981; Edwards & Scullion, 1982). Several authors have noted that arrangements that may appear acceptable in the periods of relative stability or gradual growth, may hinder significantly managerial freedoms of innovation at times of crisis or rapid change. And as some recent developments indicate, in periods of crisis, managements may have become increasingly tempted towards seeking a reduction in the scope and impact of traditional collective bargaining arrangements.

But in addition to such "macro-level" changes, it is also important to try and understand shopfloor behaviour in the context of the division of managerial labour and the tensions and contradictions thereby involved. To give the obvious example, it is perfectly consistent to see the proliferation of job controls as being appropriate to the managerial "sub-goal" of continued and smooth production, maintained in a period of relative prosperity, nevertheless having significantly damaging consequences for longer-term managerial goals of rates of profitability and control over the labour process. Thus to seek to characterise collective bargaining and its outputs as either simply co-operative or antagonistic is misleading. It contains elements of both, and tendencies in one direction or another have to be understood in the context of changing complex situations.

The State and the Encouragement of Co-operation

There is little point in pretending that the policies of the state have had anything but deeply contradictory implications for the relationships between unions and employers. Different incomes policies, wider economic goals, and policies on industrial relations practice, to take only the three most obvious, have been internally inconsistent and mutually contradictory. But having made that general disclaimer there are a few general patterns that may be discerned.

The first is the long-established interest in seeking to promote procedural formality and correctness. The activities of the NBPI, the CIR, and ACAS over the years all testify to this, as do certain periods of incomes policy, and bits of legislation since the late 1960s. As has already been suggested in this paper, such procedural innovation may have, deliberately or unconsciously, the effect of developing leadership, hierarchy and bureaucracy within shopfloor organisations and hence also, the possibility of shared definitions of problems and practices.

Second, specific legislative acts have sought to remove from "bargaining" issues which might pose problems for both managers and union officials if they were to be handled through collective bargaining machinery. Perhaps the classic case in this respect was the 1965 Redundancy Payments Act which, it has been argued, had the effect both of legitimising the notion of redundancy, and shifting attention in redundancy away from the loss of jobs towards financial compensation, allowing greater scope for compromise. Modern "bargaining" over redundancy tends to concentrate on how much more than the statutory minimum compensation can be achieved. In an analogous manner, perhaps, the unfair dismissal legislation facilitates the avoidance of potentially intractable problems of principle again by raising the possibility of equating jobs with cash. And of course both pieces of legislation can allow stewards and managers to avoid the need for negotiating altogether, by allowing both to fall back on the legal provisions. Thus a steward can now avoid embarrassment that may be incurred by defending an "impossible" dismissal case, by referring the matter to the legislative apparatus.

But not all attempts by the state to "smooth"

industrial relations by encouraging an extension of "co-operation" have been well received. Most obviously, the employers' reactions to the Bullock proposals for industrial democracy indicated clear limits to employers' interest in such "co-operation". More recently the employers' hostile reaction to the EEC "Vredeling" proposals for information disclosure has confirmed their unpreparedness to accept legalistic regulation of the nature of management-union relations. Employers wish to be able to control the content of consultation. But EEC proposals are having an impact. As the Engineering Employers' Federation recently noted, the government "would resist damaging EEC proposals - provided its position in the Council of Ministers was not undermined by a failure of British employers to implement voluntary employee involvement measures". Thus the noises coming from Europe constitute one of the forces at work in the rapid recent growth of joint consultation, a development noted in several recent works (Dowling et al., 1981; Cressey, et al., 1981; Daniel & Millward, 1983).

The Reappearance of Joint Consultation

As suggested earlier, the essentially co-operative basis of collective bargaining may serve certain managerial purposes well enough at times. However, the acceptance of collective bargaining as the appropriate method of internal regulation may have costs for management. In particular, collective bargaining may build significant delay into the introduction and implementation of change, especially if it occurs when employers consider themselves unable to facilitate the change through a "buy-out" (i.e., financial compensation for the acceptance of change). There is some evidence that during the middle and late 1970s some employers came to consider the delays and costs of collective bargaining as economically damaging, and sought ways of reducing its impact, a process facilitated by the weakening of trade union bargaining power through the effects of economic recession. In particular, managers seem to have sought to redefine in some ways the nature and scope of their relationship with the trade unions within their companies. One clear sign of this has been the recent growth in joint consultation.

Since the late 1970s committees explicitly

concerned with consultation have reappeared in significant numbers. They are not, however, concerned to address themselves to relatively minor matters - such as canteen facilities - which were often the staple diet of earlier attempts at joint consultation. These new committees occupy themselves rather with the key strategic decisions of the plant or company. Investment plans, the development of new products, problems of productivity and competitiveness, national and international, amalgamation, takeover and rationalisation are the agenda items of the new committees.

As with many other recent developments, the motives for the introduction of these new structures are mixed. Pressure for greater worker and union involvement in such strategic decisions has been increasing since the mid-1970s, partly as a consequence of EEC decisions, coming to a head in the Bullock industrial democracy debate of 1976-8. Thus Cressey and his colleagues suggest that joint consultation was seen by many senior managers as a preferable alternative to other proposals for increased participation emerging from the EEC, Westminster, and elsewhere. Both Cressey *et al* and Dowling *et al* identify management as the main instigator of these schemes. Despite ambiguities and uncertainties in managerial views of the precise reasons for the introduction of such schemes they were, according to Dowling *et al*., unanimous in the view that consultation or participation represented to management a preferred alternative to collective bargaining, since it unambiguously enshrined the right of management to take decisions unilaterally.

Thus one intention of managerial moves towards increased joint consultation seems to be to reduce the influence wielded by shop stewards through collective bargaining. Part of the strategy is an intention to involve stewards (and through them the workforce) more closely in an understanding of the problems and issues confronting the company and hence of the logic and inescapability of the conclusions and policies proposed by management. But it is important to note that the logic of this strategy rests upon the maintenance of the representative structure of the workforce, and of its authority and legitimacy, rather than upon their destruction. It is no use convincing worker representatives of the validity of managerial policies if those same representatives carry little weight with their constituents.

Managements are therefore playing a balancing

act between weakening aspects of the union role on the one hand while maintaining a degree of bargaining credibility on the other. What the long-term consequences of such a strategy might be in the event of further changes in the economic environment surrounding management/union relationships remains to be seen.

Conclusion

This note has attempted to sketch out the basis for co-operation between managers and shop stewards within the framework of the institutions. In doing this it has drawn attention to the apparent convergence of interests - substantive and procedural - that may exist between (some) managers and (some) union representatives. Following this an attempt has been made to undermine crude distinctions between "consultation" and "negotiation" especially where the latter is seen as reflecting primarily antagonistic relationships. Instead perhaps they are better seen as processes which shade into one another, both based in a strong routine co-operation, achieving different significance with changing contexts. Such co-operation is not without costs, however, for both management and workers, and from time to time attempts are made by either side (sometimes also with a degree of state involvement) to re-cast the terms of the co-operative relationship.

REFERENCES

Batstone, E.V., I. Boraston and S. Frenkel, 1977: Shop Stewards in Action, Oxford.

Brown, W., 1973: Piecework Bargaining, London.

Brown, W., ed., 1981: The Changing Contours of British Industrial Relations, Oxford.

Cressey, P., et al., 1981: Industrial Democracy and Participation: a Scottish Survey, Research Paper No. 28, London: Department of Employment.

Daniel, W.W. and N. Millward 1983: Workplace Industrial Relations. London.

Dowling, N.J., 1981: Employee Participation: Practice and Attitudes in North-West Manufacturing Industry. Research Paper No. 27, London: Department of Employment.

Edwards, P.K. and H. Scullion, 1982: The Social Organisation of Industrial Conflict. Oxford.

Fox, A., 1973: "Industrial Relations: a Social

Critique of Pluralist Ideology", in J: Child (ed), Man and Organisation. London.
Hyman, R., 1971: The Workers' Union. Oxford.
Hyman, R. and T. Elger, 1981: "Job Controls, Employers' Offensive and Alternative Strategies", in: Capital and Class, 15.
Marchington, M. and R. Armstrong, 1983: "Shop Steward Organisation and Joint Consultation", in: Personnel Review, XII, No. 1.
Ogden, S., 1981: "The Reform of Collective Bargaining: a Managerial Revolution?", in: Industrial Relations Journal, XII, 5.
Sisson, K., 1984: The Management of Collective Bargaining: an International Comparison Oxford.
Storey, J., 1983: Managerial Prerogative and the Question of Control. London.
Terry, M., 1983: "Shop Stewards Through Expansion and Recession", in: Industrial Relations Journal, Autumn.
Turner, H.A., G. Clack and G. Roberts, 1967: Labour Relations in the Motor Industry. London.

Chapter Nine

SOME CURRENT STRATEGY PROBLEMS OF THE ITALIAN TRADE UNIONS*

Hartwig Heine

Strategy Problems and Signs of Crisis in the Italian Unions

The current political and economic situation confronts the Italian trade unions with three specific problems in the field of strategy:

Union Policies are beset by the tensions between two sets of goals. Firstly there is the heritage of "egalitarianism" (reduction of wage differentials and separate wage-bargaining groups, collective upgrading) and the battle for improved "quality of employment" (control of working conditions and intensity of labour, job enrichment). And, secondly, there is the union objective of economic transformation and societal control of the accumulation process - which implies union responsibility for all productivity, and is expressly presented as such by the unions.

Union Representation faces problems due to several contradictions: (1) the claim to represent the whole industrial working class versus the reality that semi-skilled workers in large firms enjoy a privileged position; (2) the claim to represent all classes and all levels of labour (e.g., including

*The following analysis is drawn from a comparative study of changes in industrial relations in selected firms in Italy and the Federal Republic of Germany. The results of the study, in which the author participated and which was funded by a grant from the Volkswagen Foundation (Stiftung Volkswagenwerk), are presented in full in the final research report.

the unemployed, the underemployed and casual labour) versus the minimal degree of representation actually extended to these marginalised groups within the working class; (3) the claim to be building a sociopolitical perspective for all classes and levels of Italian society versus the immediate interests of the union-represented base.

The Structural Organisation of the Unions and the levels at which disputes are settled within this structure are bipolar. At one pole is the very finely structured decentralisation achieved by the system of delegates and factory councils; and, at the other, is the considerable degree of concentration of competence at the top of individual unions, union confederations and the united CGIL-CSIL-UIL Federation formed in 1972 (cf. Regalia's contribution in this volume). One aspect of these tensions is the extensive mutual autonomy of the union base and the union leadership. On the one hand, at plant level in particular, Italian unions have almost unlimited power to articulate their own demands and to express dissent from "official" union lines. On the other hand, the union machinery has very effective means of blocking the transmission of democratic opinions from the base to higher levels. This is particularly clear in the practice of settling differences among the three politically aligned unions (and also between various organised "currents" within individual unions or confederations) by internal mediation at levels above the factory, rather than through involvement of the rank-and-file.

Consequently, there is overwhelming evidence of a crisis in the Italian unions:

- The level of union organisation increased steadily from the end of the 60s until it peaked in 1977; since then it has been falling.

- "Autonomous" unions are becoming more militant and are also clearly proving more attractive. These operate outside the CGIL-CISL-UIL Federation, particularly in the service setor, and usually restrict themselves to protecting very narrowly defined interests within particular professional groups.

- The factory-wide Delegates' Councils which were created in the wake of the "hot autumn" (autunno caldo) of 1969, overturning long-established traditions of union organisation, culture, and demands, have reached an acute crisis.

For they no longer adequately represent certain sections of the workforce (white-collar workers, skilled blue-collar workers and some skilled workers) and have also lost power within the factory.

- The process, which began in 1968/69, of unification of the three politically aligned union confederations, the CGIL, the CISL and the UIL, is now in retreat. There are violent disagreements within them and among them regarding the line to be taken. At the beginning of 1984, Craxi's Socialist government issued a decree restricting the automatic indexation of wages (scala mobile). This caused a rift in the unions: the CISL and UIL stood against the largest confederation, the CGIL, and, within the CGIL, the communist majority opposed the socialist minority. For the first time in 15 years, the leaders of the three trade union confederations found themselves unable to call a joint rally on May 1st.

- Since the "March of the 40,000" (cf. contribution by Matteo Rollier in this volume), there is a real possibility that Italian capital might succeed in organising a mass movement <u>against</u> the unions.

Egalitarianism and the Concept of an Alternative to Societal Transformation

a) The demand for egalitarianism became dominant in the labour movement in 1968/69. Its main supporters were the semi-skilled workforce (the "mass labour force"), and the result was soaring strike rates and great union victories. The latter included: shop by shop election of union delegates and the creation of factory-wide Delegates' Councils; active consensus between union leadership and rank-and-file membership which persisted until the mid-70s; and effective control by the workforce and factory councils over all aspects of the use of their labour (leading to so-called "rigidity" rather than flexibility in its deployment by management). "In the space of a few years, Italy changed from the country with the weakest trade union organisation in the Western democratic world (as regards plant presence, unity and degree of organisation) to the country which perhaps has the strongest trade unions of all" (Rieser, 1981a: 165).

This growth in union power was based in part on the expansion of the mass labour force and its apparent confirmation of the egalitarian assumption that the working class is (or would become) homogeneous. The inadequacies and failings of this assumption are clear today (cf. Accornero below). In the long term it alienated blue-collar skilled workers, middle-range and top-white-collar workers, and technical staff. The egalitarian wages policy compressed wage differentials to such an extent that the only way technical staff and white collar workers could realise their greater market value was to go outside collective pay agreements and make bonus deals over whose payment management had sole control. Paradoxically, the pursuit of egalitarianism caused further problems in just those factories where the unions were most successful in demanding a continual process of collective upgrading, affecting not only pay but also skill levels and the organisation of work. The reduced number of wage levels and the convergence between them became an obstacle to further progress in this area.

Egalitarianism presupposes control "from below" by a homogeneous group of workers. Thus it is powerless in the face of such managerial avoidance strategies as greater decentralisation of production (production by small and medium-sized factories, transfer of production to other countries). Indeed all forms of modernisation of the production apparatus which take direct control of the production process out of the hands of the mass labour force (e.g., introduction of computerised central process control) undermine egalitarianism. These avoidance strategies have prepared the ground for head-on attacks by management on union-imposed "rigidity" in the labour process and on union organisation in certain vitally important firms (e.g., Fiat, cf. contribution by Rollier in this volume).

Egalitarianism has also proved powerless against inflation and economic crisis. Likewise it has proved too narrow to satisfy "new demands" such as the search for an improved quality of life and work, active involvement in cultural production, and individual self-determination. This has resulted in a marked rift between the "generation of '69" and the younger workers (Trentin, 1980; Rieser, 1981 b).

It was argued above that egalitarianism cannot represent the elementary interests of skilled blue-collar workers, technicians and white-collar workers. Nor can it "hegemonise" the weaker ranks of the working population (employees in small and medium-sized firms, casual labour, home workers, the mass of unemployed, and particularly unemployed youth) and other groups of the population whose support must be won if a social block with a majority voice is to be formed (farmers, young people, women's movement).

b) In their study of trade unions in the Federal Republic of Germany, Bergmann, Jacobi and Muller-Jentsch distinguish between "co-operative and conflictual" organisational strategies (1975). They identify the West German trade unions as operating on the "co-operative" model and the British unions as operating on the "conflictual" model. Since conflictually orientated unions by definition refuse to accept responsibility for the economy as a whole, this term cannot be applied to the Italian trade unions. For, although they are conflictually-orientated like the British rather than West German unions, they also claim even more "responsibility for the economy as a whole" than the latter do.

This attitude has been strong in Italian unions since the end of the fascist era. It has two main roots: firstly, the distinctive and uneven development of the Italian economy (particularly the North-South problem) and, secondly, Gramsci's concept of "hegemony" as interpreted by the Italian left, according to which the labour movement will become the major driving force and the guarantor of the development of national forces of production.

As trade union battles intensified in 1968/69, the Italian unions found themselves in a new situation. So extensive were the battles they fought, so dramatic the results, and so high the level of social power they gained, that their "compatibility" with the nation's socioeconomic system was no longer a problem which concerned only union theoreticians. It became a tangible political problem. The unions' "project" of socioeconomic transformation now had to incorporate yet another factor: rendering the socioeconomic system compatible with the unions' own mode of operating and the positions gained thereby. In other words, they had to consider how to structure the interaction of these two factors. Indeed, one of the greatest problems faced by the Italian unions since at the very latest 1973/74 (when uncontrollable inflation and prolonged economic crisis began), has been the incompatibility of the socioeconomic environment and union strategies. For does not the very "progressiveness" of the positions gained tend to rebound on the unions which fought so hard for them? (Rieser, 1981 a). The changes in union strategy since the beginning of the 70s can largely be interpreted as a response to this increasingly acute problem.

Four Phases of Union Strategy

The first phase, from 1968 to 1972/73, is characterised by the absence of an economic crisis. The problem of compatibility was no more than an abstract response to the propagandist predictions of gloom uttered by Italian employers. The revolutionary impetus which had gripped the labour movement and fundamentally altered the balance of power in the factories aimed at a substantial social transformation. The theme of a press conference given by Novella, the General Secretary of the CGIL, at the beginning of 1970, was "Take the demands of the workers into the heart of society". In Novella's words,

> if union achievements are to be consistent and enduring, they must be transformed into our objectives of reform and power, ... As long as the unions are regarded as a means of rationalisation and are therefore expected to behave so as to avoid conflict, we are creating an illusion of reconcilability which just does not exist ... The unions have no choice but to hasten transformation and break through the current stalemate. Further, the unions must develop a very high level of expertise in order to release the alternative potential of the workforce (Novella, 1970).

The sociopolitical reforms demanded, beginning with themes of pensions, abolition of wage zones, taxation, housing, and later adding the demand for investment in the South, were the symbolic first breaches in the ramparts of the existing social structure. Union demands, whether directed at capital or the state, "can only be rooted in a project of social revolution which we regard as imminent" (Perulli, 1981: 38).

The second phase, from about 1973/74 to 1976, was the initial response to the advancing economic crisis and soaring inflation. The "Rimini Strategy" (after a central conference of factory delegates in Rimini in April 1974) was an attempt to remain on the offensive, and, even during the crisis, to consolidate further the advances already achieved. In this way it was hoped to disengage the mechanism of capitalist recession and to implement the planned development of national resources. The unions' strategy centred on company investment policies with the intention of forcing a coherent social investment policy. First individual agreements with companies, then the 1976 national wage agreements, obliged management to provide the unions with a whole system of information on current and planned investments. Thus equipped, the unions intended to develop alternative investment programmes with new investment priorities.

According to the Rimini strategy, the state was to create the necessary environment by preparing plans for the various sectors of industry, taking into account both investment requirements and sales forecasts. The scope of the whole project far exceeded that of the West German concept of co-determination (Mitbestimmung), and, at the same time, set out to maintain and even extend the conflictual, autonomous nature of trade union organisation. The theoretical problem of compatibility had become only too real with the advent of the economic crisis. Now it was to be solved by direct union influence on the prevailing economic environment, by suspending such "natural" laws of the economy as the refusal to invest when returns are falling. That in essence was the so-called "titanism" of the Rimini strategy.

What undermined this strategy was its failure to achieve much in the way of real results. The unions certainly gained important rights of access to information on industrial investment policies. But this did not help them gain the objective they desired: implementation of a new type of anticyclical investment policy, a new development model for the Italian economy. The crisis deepened and so did the problem of compatibility.

It is no mere coincidence that the third phase, which lasted from about 1977 to 1979, is the period in which the country was governed by a coalition of "national solidarity". Although it was not a member of the government, the PCI helped to formulate policies. Union policies began to take a new line, formulated in the Strategy of Eur (named after the central conference of delegates held at Eur, outside Rome, in February 1978). No longer were the unions' gains and continuing demands to be regarded as independent variables while the economic system was regarded as a dependent variable. The interplay of the two was now recognised. Further, an "alternative model" was to be introduced, and a "productive solution" found to the crisis. First of all, however, it was recognised that in the interests of this objective,the "rigidities" which had been fought for in the past, and the new demands now being made, could be modified within certain limits, or used for bargaining purposes. The unions would now "sacrifice" some of their own objectives quite consciously, in order to give them bargaining power: "If we are to emerge from this crisis ...... we have to understand that we are talking about a bargaining process, where every political concession we make must be matched by a corresponding political gain" (CGIL-CISL-UIL, 1977: 30). The state was addressed more directly than ever before, an attitude which expressed the implicit faith of the unions in the government of national solidarity (Rieser, 1981 b). The unions called for a long-term policy of planned restructuring, selective promotion of investment, promotion of full employment, a regional development policy for the South, tax reforms in favour of those with low incomes, and reform and rationalisation of public spending. In return, the unions were prepared to "concede" the containment of labour costs by keeping wage demands consistent with the general development objectives, and to grant increased flexibility of labour (mobility within and between companies, overtime, self-control with regard to absenteeism, holidays, introduction of shift work, etc.).

It is important to emphasise that these were not concessions in the sense of deserting hard-won ramparts and handing them over to the class enemy in the hope that he would make concessions himself in other areas. The Strategy of Eur and the "sacrifices" implied therein are interpreted quite differently by union leaders such as Trentin. Indeed, by making these sacrifices, the labour movement is now living up to the principle of compatibility and is thus proving its claim to be a hegemonic force and the motor of the social production process. Trentin made this quite clear when he explained why he was in favour of less "rigidity" in labour mobility:

> Controlled, collective mobility of labour ... presupposes the political realisation within the labour movement that, particularly on a strategic level, a purely defensive battle for jobs everywhere, regardless of the situation, is

> bound to fail. For this defensive attitude becomes self-contradictory and is a true sign of regression if, at the same time, the labour movement generally and knowingly supports a policy of political restructuring. But it is also true that the decision to control labour mobility requires the conquest of political power and that the unions have not yet achieved this. The power in question is the power of control over regional movement of labour, job placement and training and retraining programmes, in other words, qualitative reprogramming of the labour force, and a labour policy actively controlled by the working class (Trentin, 1980: 187 f).

The Strategy of Eur also failed to achieve its bold goals. Despite internal problems with the base, the unions delivered their part of the bargain. Wage demands and absenteeism were reduced, local conflicts became less frequent, and demands at company level became more "qualitative". The mobility concessions were granted and the unions also accepted a revision of the system of calculating the automatic wage escalator (scala mobile). But the state did not keep its side of the bargain. Parliament threw out some legislative proposals (e.g., the law on pensions reform), and the state bureaucracy immediately made a mockery of the original purpose of those laws which did get through parliament (e.g., a law on the restructuring of industry). The state thus proved that it could act as the political guarantor for the compensatory measures proposed as <u>long-term</u> counterparts to immediate union concessions. The end of the coalition of national solidarity and the return of the PCI to the opposition in 1979 indicated the fundamental precariousness of union-party alliances.

The attempt of the Strategy of Eur to solve the problem of compatibility thus broke down because of the irreversible distinction between its two components: the immediately negotiable objectives traditionally pursued in union bargaining (wages, hours, rigidity), and the ever-receding compensatory demand for qualitative socioeconomic transformation.

The <u>fourth phase</u> began in 1980. It is defined by diminishing success in formulating uniform union strategies towards industry and the state. Thus this phase is one of <u>union crisis</u>.

The internal weakness of the unions is growing to a frightening level. The crisis of union representation is becoming more acute. There are two reasons for this: first there are the problems of representation resulting from the union doctrine of egalitarianism of 1969, which affect all those groups within the workforce which do <u>not</u> form part of the unionised semi-skilled workers in the large factories in the North. These problems are compounded by the second type of representation problem, born out of the failure of the Strategy of Eur, now affecting the core of the unions. Since 1980 there has been a growing policy of blocking political bargaining with the government and industry. Union concessions granted in the "national solidarity" phase are seen by the union members themselves as a sign of weakness or even "betrayal". With the return of the PCI to the opposition, different political currents threaten to increase the splits in the unions.

Without a doubt, the initiative has passed into the hands of management. A phase of fundamental industrial restructuring has commenced, resulting in partial or complete shutdowns in many divisions of industry. Hundreds of thousands of employees are being laid off, primarily by means of permanent reduction in hours. In certain privately owned key industries (e.g., Fiat), management has launched a head-on attack on the existing structure of industrial relations. The goal goes beyond simple suppression of union control over the use of labour (thus also attacking the "rigidity" principle). It also aims to undermine union positions of power inside and outside the factory. In the middle of 1982, the Italian federation of private industries, Confindustria, abrogated the agreement on the scala mobile. Confindustria wanted to reduce the degree to which the scala mobile automatically makes up for inflation, to delay conclusion of the pending wage agreements, and to force "regulation" of disputes which occur at company level.

The three union confederations found it very difficult to reach anything like a common defence against this onslaught from private industry. But they did manage to force government intervention. The conflict ended in January 1983 with a trilateral agreement between the trade union federations, the industrial federations, and the government. This amounted to a compromise or "cease-fire" for all parties. The unions accepted a 20% restriction of the automatic wages indexing mechanism, restriction of contractual wage increases to a level which - combined wit the scala mobile - would keep effective wage increases within the limits of the "planned inflation rate, suspension of pay discussions at plant level for 18 months, and partial liberalisation of the hiring of new workers (which had always been very strictly controlled by the unions). The federation of industries agreed to no further revision of the automatic indexation system and to no further regulation of company-level disputes, and said they would no longer block conclusion of the wage settlements. The government agreed to meet part of the costs incurred by both parties as a result of the concessions they made. The unions were promised tax relief for the lower and middle income brackets by means of restricted tax progression, improved child benefit and containment of increases in prices and duties in the public sector to the "planned" inflation rate. Industry was guaranteed fiscalisation of a proportion of the employer's social security contributions.

Subsequent developments made it only too clear that these agreements were more in the nature of an interim "cease-fire" than the beginning of a long-term, trilateral "concerted action". Internal union disagreement on the policy line became more intense, and the union confederations could no longer be relied on. The industrial federations, meanwhile, continued to block wage settlements and to dismantle union positions. The state only kept its word on the immediate promises. Clearly the unions were now on the defensive, having accepted the loss of strategic ground. Although the agreement guaranteed the real income of wage-earners with regular employment contracts, and in a more diluted form also protected "the consumer", it offered neither a long- nor a short-term perspective in the form of a new production and employment policy which would also protect the interests of the unemployed, the underemployed and other disadvantaged groups. In addition the wages policy accepted in the agreement further weakened the plant-level union base (factory councils) by restricting their bargaining scope.

In the absence of a common union political line, the Craxi government's move to suspend the scala mobile for 1984 caused an open rift between the unions. The government decree justified the suspension as part of the battle against inflation, and also reinforced it with a number of compensatory tax measures. The two union confederations closest to the government, the CISL and the UIL, together with the socialist minority within the CGIL, accepted this. The communist majority group rejected it - in part because of their political distance from the government. They emphasised the injustices of the Italian tax system, and also expressed their fear that this government measure would be taken as a precedent in future years. In this case, unions would lose a major instrument for controlling state social and economic policies. Further opposition to the decree came through an unexpectedly strong mobilisation "from below". The protagonists here were the factory councils. These had already been pronounced lifeless by many observers but they still helped to stage a mass demonstration of more than 700,000 in Rome at the end of March 1984. A combination of extraparliamentary action and parliamentary filibustering destroyed the decree as it stood. A second, considerably watered-down version was then introduced and finally became law in June 1984.

Discussion of Solutions to the Crisis of the Unions

a) Various solutions have been suggested to the problems facing the unions. The left union opposition, supported by activists in several large factories, insists on a "hard" strategy. This would involve demands for the employees of large companies, consciously ignoring the problem of societal compatibility and disregarding unrepresented elements of the working class. Although this approach is often presented with the slogan "Back to the line of '68", it actually differs from the original line in no longer laying a claim to social hegemony. Union policies have only one objective: the protection of union members. Social mediation and synthesis are beyond their brief. This approach may appear very inviting in terms of the absolutely obvious symptoms of crisis in the trade union movement in Italy. But it is also totally without perspective. To restrict the issue to the defence of "rigidity" and to disclaim all responsibility for the economic crisis would be to condemn the unions to passivity towards a trend which threatens to undermine the positions they fought so hard for in the 70s.

A similar option is implied in the variation proposed by the "right". In particular the UIL, the smallest confederation within the CGIL-CISL-UIL union federation, puts forward three objectives. It calls for the "modernisation" of the Italian unions on the West German model, i.e. it is in favour of a clear separation of union interests and "general" policies. It calls for "pluralistic" representation of interests. And it calls for a "social pact" which recognises the inevitable incompatibilities of the existing socioeconomic order. It seems very unlikely that this concept could be accepted by the Italian trade union movement. Even if the political culture of the Italian labour movement did not bluntly contradict such a solution, it would be quite impossible to realise in the present political situation.

The solution suggested by the CISL is also similar. The CISL hopes to overcome the union crisis by confronting the new liberalism with a form of concerted economic steering together with the formation of union "accumulation funds" to support self-governing factories, co-operatives, etc. But its desire for political bargaining with the state and capital assumes that the Italian government can act as political guarantor. It also ignores the inevitable consequences such concerted economic steering would have for the representation problems of the unions. For localised negotiations would have to be more closely controlled by central bodies, or even cease completely. The employment problem is primarily understood in terms of distribution of the hours available, and the CISL supports very strongly an overall reduction in the working week. This conception is far too limited.

b) The CGIL has a far more complex view of future union strategies. On the one hand it wants the unions to retain their autonomous voice over the evolution of the national economy. On the other hand, while carefully taking into account the "plurality" of the interests and requirements represented, it also wants to extend the scope of union representation. The CGIL has set itself "the ambitious goal of reunifying the labour world" (Lama, 1981). Three methods are proposed to offset the privileged position of the semi-skilled industrial workers:

- previously "underprivileged" sectors of the workforce, particularly technical and clerical staff and skilled blue-collar workers, should be better represented in union policies and demands and more fully integrated into union institutions. The compression of wage differentials should be reversed and the wage span broadened again. Further, the demand was made at a conference on wages reform for a return to a percentage basis for operation of the automatic wages escalator. In the second half of the 70s the unions pushed through a ruling that, in order to protect the lower income brackets, the amount of inflation compensation paid should be the same across the board. This proved to be a major force in the compression of wage differentials. Now it is to be discarded.

- "Marginalised" sectors of the workforce (the unemployed, the underemployed, part-time and home workers) are to be represented by the unions. The hope has been expressed that these groups might become the new protagonists, as happened with the mass of semi-skilled workers after 1969 (Trentin, 1981).

- Groups with "new demands", particularly youth and women, should be more strongly represented in union policies. This is because "to a certain extent they call into question the forms and characteristics of the central role of the employed working class, which persists in spite of everything" (Lama, 1981).

As far as the content of union demands is concerned, the CGIL subscribes to an offensive approach to sociopolitical transformation. "A strategy aspiring to reunification of the labour force has to seek unity and synthesis at the highest level: in a project of societal transformation ... In the current situation, the objective of class unity cannot be divorced from the project of transforming societal organisational structures" (Lama, 1981). Lama objects to the desertion of sociopolitical objectives which is implied in the proposals of both the left and the right.

> We have a concept of the role of the unions which pays heed to the needs of the country as a whole and not simply to the needs of the class we represent ... If we cannot reach solid ground in economic policies, if we do not put all our effort into rationalisation and revival of our production apparatus, then we inevitably undermine our ability to maintain real wage levels. And this process has already begun (Lama, 1981).

The Italian unions were already experiencing problems in the 70s of reconciling this reform "at the highest level" with the immediate interests of the union base. At the beginning of the 80s, union leaders stated their belief that a "qualitative cultural leap" (Trentin, 1980) was needed in the consciousness of their members in this respect. The fact that this did not occur is one of the reasons given for the current "delegate crisis". The shift in labour organisation in the factories to provide consistent job enrichment is seen as the decisive link between (a) immediate union demands regarding wage levels and working conditions, and (b) more general, sociopolitical objectives. Failure to raise this issue is now seen to account for the absence of a "qualitative leap" in union strength at the end of the 70s. It has now been adopted as the key to the CGIL's proposed strategy. Thus changes in labour organisation (healthier working conditions, increased content and purpose, dismantling of hierarchical structures) should be raised to the status of "national policies". State support is only to be given to industrial restructuring projects which involve the desired type of reorganisation of work. This policy is designed to guarantee that the employees' interests are more fully taken into account in the current restructuring process.

Improved quality of work takes on a key role with the inclusion of groups with "new demands" in active union policies.

> Not only can the battle for a new quality of work produce tangible results. It is also necessary, even decisive, if the forces divided by the labour market are to be reunited for a "humane" project, i.e. a project in which the new demands for freedom which initially grow up <u>during</u> work can be fulfilled <u>via</u> work, labour and struggle (Trentin, 1980: 233).

This is a decisive stance in opposition to those dualistic views (cf. Gorz, 1983) which see an irreconcilable rift between the industrial realm of necessity and the private, non-industrial realms of freedom, a rift which also divides society.

In the view of the CGIL, control over the reorganisation of work simultaneously bestows control over the means to secure an increase in industrial productivity (which it accepts as necessary). This does not occur through union capitulation, but by establishing hegemonial positions within the plant. Non-Italian observers find it strange to hear the leaders of the CGIL demanding the right to solve the productivity problem themselves:

> Our proposed treatment of the condition is to increase Italy's industrial productivity, to raise the level of professionalism and to mobilise all our reserves in the interest of improved productivity. The specifically communist element in my statement about restructuring the economy is that we truly desire it, whereas the others only talk about it. But we also want to be involved in controlling it (Lama, 1984).

The CGIL believes that the first step is to abolish assembly lines and to introduce labour groups which are largely self-regulating.

> A way of increasing productivity and creating new resources for accumulation can be born out of new forms of labour organisation, a new quality of labour and new forms of professionalism, the creation of collective responsibility of the producers, and a realistic degree of autonomy at work. Deciding to increase productivity does not mean making things easier for the capitalists, but quite the reverse: an alternative is created to the tendency associated with crisis, which is to found efficiency on the reintroduction of classical forms of exploitation (Garavini, 1981).

It remains to be seen, however, whether the forceful new approach to the reorganisation of work really lives up to the vision of the CGIL. There were structural reasons for the failure of this approach in the late 70s, following its key role in the union policies of the earlier years of the decade. A number of studies (e.g., Chiaromonte, 1976; Heine, 1983) suggest that circumstances must be very specific before a company gives in to union pressures for a substantial reorganisation of the work process. The few successful examples which can be quoted cannot be generalised to the extent which the CGIL would like.

The CGIL has also developed a detailed view of the <u>institutional means</u> required for the unions to bring about "democratic programming" of the accumulation process. This is a response to the poor achievements of the unions' attempts at social reform over the past decade, with regard to both levels of intervention (company, sector, state) and to mobilisation of their own base.

The state is to become a more important

instrument of "democratic programming" than ever before. The keys to "democratic programming" were previously in the hands of uncontrollable central bureaucrats. Now they are to be decentralised and passed into democratic control. Regional and communal centres of autonomous decision making will be set up. They will have the right to make material decisions on questions of territorial economic planning, credits, control of investment, labour market policy, mobility, and professional training. These measures require state reform. The unions are particularly keen for an active role in these decentralised decision-making processes. For this will allow them to develop their own objectives covering more than just one industrial sector, to obtain recognised negotiating rights on this terrain, and to deploy all available union resources.

However, since the CGIL assumes that state-mediated control over the economy would not be sufficient for effective programming of the accumulation process under present social conditions, the second lever is to be the "company plan" (piano d'impresa) (cf. detailed legislation proposal, CGIL, 1980). This would require companies to prepare a precisely calculated plan for the coming years, with details of projected production, sales, and financial policy, and to present this to both the state planning agencies and the unions. The unions would have the right to raise objections to these plans, to use union powers to modify them, and to check that they are followed. Financial sanctions could be imposed on companies which fail to present such plans or fail to fulfil them. Knowledge of these plans, and the power to modify them, would enable the unions to implement true "programming" on a territorial level and to gain perfect control over accumulation.

The CGIL, recognising its past problems of bringing about "alternative economic development", has clearly decided to seize the offensive. In the past companies may have succeeded in keeping their policies out of the grasp of the unions. Now they are to be legally obliged to reveal them to the state and the unions, and to discuss them with these parties. In the past, any sociopolitical initiative of the unions which might have cleared the parliamentary hurdle was rendered ineffective by central bureaucratic intervention. Now the decision-making authority of the state is decentralised and becomes directly accessible to the union base. The "titanism" of the past, then, has not disappeared. On the contrary: the results of earlier attempts to control companies and the state have been critically analysed. Thus even greater and more comprehensive control over national economic and social developments is demanded by the unions - beginning with fundamental reforms of the state. This brings us to the question of the forces which will pursue these reforms.

The demonstration of the 700,000 on March 24, 1984, bore witness to the continuing high potential for mobilisation of the Italian labour movement. It must be remembered that this demonstration was in defence of a very particular asset, namely, the automatic indexation of wages. But all the studies carried out on plant level in recent years show that the frequency of "microconflicts" over plant or department-specific issues is falling. Thus it would appear that the union base is more readily mobilised for causes which go beyond the plant - although in the last round of wage agreements, which in the case of the metal workers dragged on for 18 months, this was put to a severe test.

The plant-level industrial relations structure installed in the large and medium-sized factories of Northern Italy at the beginning of the 70s is now clearly wearing thin. In those factories (not Fiat for example) where the fabric is still more or less intact, negotiations have been reduced to a ritual with very little material outcome. In the case of pay negotiations this is because they have become increasingly centralised. Moreover, wage differentials have been compressed to such a degree that negotiations on collective upgrading, which were the main feature of plant-level representation in the 70s, are losing all relevance. The problems of the factory councils in representing growing sections of the workforce (white-collar workers) are obvious. The close relationship between delegates and workforce, which was the real strength of the factory councils, is crumbling.

One means the factory councils employ to make up for their loss of autonomous resources is the formation of alliances with particularly "accessible" management levels in an attempt to preserve their political status within the factory. But this puts them under heavy pressure to professionalise. They have to become experts in work organisation, technology and production planning, etc. This only serves to increase the distance between council delegates and the rest of the workforce. There are practically no relevant models for the factory councils to follow during these changes.

The development of "alternatives" to industrial policies has become a standard feature of Italian trade union culture. With the economic crisis, however, this claim to provide alternatives has lost all substance. This is particularly true at company level, despite the fact that falling payrolls in almost all factories would be expected to generate enthusiasm for such alternative policies. Can a project intended to provide a "productive way out of the crisis" for society as a whole carry conviction if it is already failing at company level and its content remains uncertain at levels above this? This seems doubtful.

For the CGIL, modification of the organisation of work is meant to be a key strategy. But there appears to be a general loss of initiative on the part of the factory councils in recent years as a result of the extremely rapid technical and organisational restructuring which has taken place. This is particularly true because the restructuring processes frequently call for a type of intervention which can no longer take place at individual department level, the level of the "homogenous group". The best the factory councils can achieve is to preserve a few principles of solidarity (e.g. collective bonuses) when presented with new forms of labour organisation. But at plant level, it does not seem that the topic of labour organisation is about to become the launching pad for a great new thrust in union policy. The deliberations on strategy put forward by the CGIL at present resemble the battle plans for a new offensive in which all the force is concentrated in the defensive.

REFERENCES

Bergmann, J., Jacobi, O., Mueller-Jentsch, W. 1975: Gewerkschaften in der Bundesrepublik, Frankfurt-Koeln.

CGIL 1980: Vorschlag der CGIL fuer die industrielle Demokratie. IRES-CGIL: Democrazia industriale: idee e materiali, Rome.

CGIL/CISL/UIL 1977: Eroeffnungsreferat zur Konferenz von Eur. In: Rassegna Sindacale, Nr. 1/1977.

Chiaromonte, N. 1976: Come si ristruttura la fabbrica. In: I consigli, Nr. 20-21, Februar 1976.

Garavini, S. 1981: Referat auf dem CGIL-Kongress vom

November 1981, in: Rassegna Sindacale, Nr. 43/1981.
Heine, H. 1983. Am Beispiel Olivetti: Das Konzept der "Anreicherung" der Arbeit in der italienischen Gewerkschaftsbewegung der 7 0er Jahre, Bielefeld.
Gorz, A. 1983: Les chemins du paradis, Paris.
Lama, L. 1981: Referat auf dem CGIL-Kongress vom November 1981, in: Rassegna Sindacale, Nr. 43/1981.
Lama, L. 1984: Gespraech mit dem SPIEGEL in Spiegel Nr. 15/1984, S. 128 ff.
Novella, A. 1970: Erklaerung auf einer Pressekonferenz Anfang 1970. Cited in: Rassegna Sindacale, a. XVI, Nr. 180, 5. Januar 1970; also in: Quaderni di Rassegna Sindacale, Nr. 51/1974, S. 44 ff.
Perulli, P. 1981: Sindacato e rivendicazione. In: Laboratorio Politico, Nr. 4/1981. Rome.
Rieser, V. 1981a: I due vincoli dell'azione sindacale. In: Prospettiva Sindacale, Juni 1981.
Rieser, V. 1981b: Sindacato e composizione di classe. In: Laboratorio Politico, Nr. 4/1981. Rome.
Trentin, B. 1980: Il sindacato dei consigli, Rome.
Trentin, B. 1981: Rede vor dem CGIL-Kongress vom November 1981. In: Rassegna Sindacale, Nr. 43/1981.

Chapter Ten

CENTRALISATION OR DECENTRALISATION? AN ANALYSIS OF ORGANISATIONAL CHANGES IN THE ITALIAN TRADE UNION MOVEMENT AT A TIME OF CRISIS

Ida Regalia

In times of crisis, what is the most convenient organisational structure for a trade union, assuming that one of its general objectives is to participate in the process of economic policy making to the advantage of the workers it represents? Is a capillary and decentralised type of union structure more appropriate? This type would allow the union to absorb all the grass-roots demands. Or is a centralised form of organisation more appropriate, which would simplify decision-making at the top without hindrances and delays resulting from too much lower-level participation?

Without doubt, most observers of trade union behaviour would argue that the second option, a centralised structure, would be the more appropriate. A centralised structure seems more efficient, reducing the costs of aggregation and mediation of all the different demands and interests dispersed throughout the industrial relations system (Streeck, 1981; Visser, 1983). Indeed, the existence of a strong unitary union movement, centrally organised, is seen by many as an indispensable requirement for a negotiated management of the economic crisis (involving agreement between unions, employers, and government), as opposed to management by government fiat, or by the free market (Salvati, 1983). The importance of a strong, centrally organised union movement for negotiation lies in the union's ability to control and co-ordinate internal tensions, involving, for instance, disagreements among different groups of workers, more or less advantaged by various policies.

Seen from this angle, the degree of centralisation and co-ordination of internal union structures has a direct impact on the ability of the unions to tackle, through agreements, the kinds of economic problems that have arisen in the late 1970s. Certainly, what really counts is not the centralisation of structures *per se*, but the fact that centralisation facilitates the aggregation and mediation of demands. This goal, it is true, can also be achieved by other means. For example, the union could encourage the development of an integrated movement through the use of symbols of unity, encouraging unitary values and discouraging the actions of minority groups of workers (Regini, 1983). In the long run, however, it is likely that this latter method will prove unreliable, or unstable, or too costly. Thus unions, negotiating with governments over important questions of economic policy, will tend in the end to modify their own internal structures if they are too dispersed and disarticulated. Their need is to make their own structures congruent with their functions in the political arena.

Accordingly, the major organisational reform launched by the unitary Federation CGIL-CISL-UIL in 1979 can be interpreted as an attempt by the union movement to rationalise its own structures. This reform has brought about wholesale changes in the internal organisation of the union, particularly among the middle level structures. The desire of the movement to reform its structures derives from its desire to intervene actively in social and economic policy-making (Lange *et al*., 1982), and the realisation that its pre-existing structures were inadequate for these purposes.

My purpose in this paper is to examine the nature of the organisational reform. Surprisingly, perhaps, the reorganisation of structures did not involve a simple process of centralisation. Logically, one might have expected such a development. Instead, the organisation was much more complex. As we shall see, it was an extremely ambitious project, which has revealed certain contradictions in the union movement, and which has had ambivalent results.

In the following sections, I will outline the major features of the structural reform, showing the effects, both expected and unexpected, which it has had on the union's capacity for effective representation and on its general operation. First, though, it is important to discuss the background to the reform. In the next section, therefore, I shall examine the main features of union organisation, which were established in the years prior to the economic crises of the middle seventies, and the progressive changes in the union's general strategy.

Italian Unions in the Mid-1970s

In the mid-1970s, the Italian union movement presented the following major structural features (1).

First, at a general level, the union had become a federation of the three major union confederations - the CGIL (Confederazione Generale Italiana del Lavoro), CISL (Confederazione Italiana dei Sindacati dei Lavoratori), and UIL (Unione Italiana del Lavoro). The three confederations came into existence with the split of the unitary CGIL at the end of the 1940s. A form of reunification came about in 1972 with the signing of a federative pact. Under the terms of the Pact, each organisation was to retain its own organisational autonomy at all levels. But alongside this existing organisation, new joint structures were to be created which would have the task of taking and co-ordinating decisions concerning the federations rather than the separate confederations.

Second, as in the past, the organisational structure of the unions continued to be characterised by both vertical and horizontal structures. The vertical structures are based on productive decisions between industries, or branches of industry. Thus each confederation is made up of a series of industrial federations. The horizontal structures are based on a territorial division. When the union movement was relatively weak and politically isolated, as it was throughout the 1950s, the horizontal structures, and, indeed, the confederal structures (which aggregated the interests of all workers at the national level) tended to be more important. With the increase in collective bargaining at the industrial and plant-level during the 1960s, the vertical structures became more important (Santi, 1983). The latter were clearly better adapted to voice and represent the distinct interests of various occupational groups.

Third, as far as regards the middle-level or intermediate structures that linked the national unions to the work-place, a crucial role was played by the provincial unions (2). Certainly, other linking structures exist, having developed in a fairly disordered manner over the post-war era. Local structures (communal and zonal), for instance, often played an important role, especially in areas of hhigh industrial density and in agricultural areas (3). From the beginning of the 1970s, regional structures have also been set up, but these have limited and ill-defined roles to play. Above all, it is the provincial unions that constitute the nerve-centre of the system, functioning as a crucial link between industrial action at the grass-roots level and decision-making at the centre. As grass-roots action increased during the 1960s and 1970s, contributing to the growth of union power, the role of the provincial union as a link became still more important. It is fair to say that in many large industrial cities, the provincial union became more important and influential than many of the national industrial federations of the union.

Finally, in the work-place, entirely new forms of representation have developed. In the early 1970s, the internal commissions (which, in fact, had never been union structures, but instead were structures of workers' representation on the German Betriebsraete model) were finally phased out. They were replaced by new unitary structures, the factory councils (known variously as consigli dei delegati, consigli di fabbrica, and consigli di azienda), which had undergone a rapid and spontaneous development, particularly in industrial sectors and some branches of the public and private tertiary sectors. Separate forms of representation which were not unitary, being based instead on union affiliation, developed in those sectors of the economy which were by-passed by the vigorous surge of union militancy at the beginning of the decade. Representatives of these structures have either been elected by union members, or nominated by the union hierarchy. Either way, such structures have a coherent organisational logic from the unions' point of view. It is ironic, therefore, that because they are present in those sectors of the economy with the least militant workforces, they play only a peripheral role in the overall strategy of the unions (Regalia, 1982a). Of less logical coherence, but far more relevant to the general development and functioning of the union movement are the factory councils. Let us examine them in greater depth.

The most important feature of the factory councils is that they are unitary and sole forms of representation in the firm. They represent both workers (who freely elect their own delegates, usually on a "write-in" system, without formal lists of candidates) and also the unitary Federation of the CGIL-CISL-UIL. The Federative Pact of 1972 stipulates, in fact, that the councils should be officially recognised as the grass-roots structures of the union. (The old internal commissions never received such reognition.) The councils, therefore, enjoy a double legitimacy since they represent both the workforce and the unions. The most extraordinary aspect of the councils is that though they are union structures, any worker has the right to become a delegate to the council, whether he is a union member or not. All that is necessary for someone to become a delegate, and therefore part of the union organisation, is to receive a majority of the votes of his work-mates in the council elections (4).

This last aspect of the councils can only be properly understood when one takes into account the very special circumstances under which councils emerged in the major industrial sectors. Three factors were relevant. First, widespread social mobilisation and a climate of hope and enthusiasm accompanied the emergence of the conflict scene of a powerful new occupational group, low and semi-skilled workers, in the 1960s and 1970s; second, it was convenient for a union movement that was organisationally weak to support, organise, and later generalise a powerful grass-roots movement. This union acceptance of the councils eventually led to far-reaching changes within the unions, culminating in the reunification of the three major confederations in 1972. The third circumstance favouring the emergence of the councils was that there were huge advantages to be obtained for workers in the decentralisation of collective bargaining to the level of the firm and the shopfloor (Regalia, 1978). The first point explains the informal character of the movement, the lack of union control of rights of representation, and the general manner in which the councils functioned. One must bear in mind that a new "collective subject", as the occupational group of low and semi-skilled workers became, would be suspicious of any forms of representation, which necessarily involve renouncing direct grass-roots action, giving instead to others the right to act and speak in the name of the group (Pizzorno, 1977). The second point explains the fact that the councils were not, and could not have been at their formation, the outcome of any constitutional agreement among the unions. Instead of firm rules laying down their role and operation, the councils are marked by their informality and flexibility. Finally, the third point explains the popularity of the councils among workers. The councils provided workers with the opportunity to freely express their preferences. They also obtained immediate and highly visible results. The very success of the councils has made them a strong symbol of all the hopes of renewal.

Overall, one can say that the union structure of the middle 1970s mirrored recent union development. The latter had been intense, tumultuous, uncoordinated, closely linked to the developments in collective bargaining, and in general marked by a strong tendency to the decentralisation of industrial relations. This decentralisation involving the growing importance of the periphery relative to the centre; of the industrial federations to the confederations; of provincial structures to central ones; and of workplace organisations to the union outside the workplace. But there has also occurred a loss of equilibrium in the union movement as some sectors of the workforce have increased their bargaining strength relative to others. Cases in point are the electrical and energy workers, as well as certain groups in the public sector. In a similar manner, there is also disequilibrium between other structures. The largest provincial unions (like Milan and Turin) and the major factory councils are disproportionately powerful compared to other middle-level and grass-roots structures.

The New Organisational Reform

Partly because of the disequilibrium in the distribution of its internal resources (and power), and partly because of the importance acquired by the decentralised structures, more orientated to the representation of particular interests, the union organisation of the mid-1970s seemed scarcely adequate to the union's strategic needs at the national level. The union's chief concern, in fact, was to be able to play a strategic role in the policy-making process, particularly in those areas of policy concerned with solving the country's economic crisis.

At the union congresses of 1977, the need for some form of internal reorganisation was affirmed by each of the three major union confederations. Two years later, at the Montesilvano Meeting of 1979, the unions acted, launching a wide-scale organisational reform.

What were the objectives of the proposed reform? Generally, the intention was to redistribute the available resources (structures, staff, financial resources etc.) more equitably and rationally. This was intended to bring them into line with the union's strategy of pursuing a more central role in the making of social and economic policy. Just the previous year, at the Eur Assembly of 1978, the union had given official approval to this strategy. This decision meant that the problem of co-ordination had to be confronted. General co-ordination would be given more emphasis than the management of details. The confederations would have to assert themselves over the various federations. Put another way, the horizontal logic would take primacy over the vertical logic. The purpose of the reform was not to proceed simply to a centralisation of the organisation at the national level. Instead, the unions underlined their intention of reinforcing the functions of coordination also at the middle levels of the organisation, increasing the powers and responsibilities of horizontal or confederal structures. A second intention was to make the channels of communications between organisation and workers more homogenous and efficient so as to make it easier for the union to find a consensus on its various strategic choices. The watch-word of the proposed reform was to build a "new confederality".

In terms of the detailed working of the union organisation, the following decisions were taken: 1. Regional structures were to be consolidated (5). Their tasks would be to direct local union initiatives and to take responsibility for links between the centre and periphery. 2. The old provincial structures were to be replaced by new local units of more restricted dimensions, based on districts (6). 3. The various federations were to be rationalised through a process of regrouping and fusions, so as to reduce their number and make them more homogenous in terms of corresponding federations in each of the three confederations. 4. The revitalisation and generalisation of the unitary structures of union and workers' representation (the factory councils) in all work places was to be carried out. Councils would be introduced in those sectors (public employment and agriculture) where they had not successfully been established, and previously existing forms of organisation would be eliminated. 5. Finally, the unions made a commitment to extend unitary structures in the community (the zonal councils) as widely as possible. The zonal councils would act as a link between the district and the workplace, providing the factory council with the framework in which to address and co-ordinate their own action.

From these proposals, it seems that the new organisation of the union would be based on the following levels: factory councils, zonal councils (facilitative), and district, regional, and national structures. The latter three levels include both horizontal or confederal elements (whose influence was to be increased), and vertical or sectoral elements (whose influence was to be diminished).

Apart from minor changes, the new organisational design, compared to the previous one, is characterised by an expansion of the intermediate levels of the organisation, resulting from the replacement of the single provincial structure by two new structures, the regional and district offices. It is also characterised by a simplification and diminution of the number of sectoral federations; and finally, by a new emphasis on a homogeneous union presence in the workplace.

Overall, the organisational reform was intended to bring coherence and order into the union movement following the years of great spontaneous growth. However, from some standpoints, the reform can be said to present surprising and disconcerting aspects. Above all, the general sense and purpose of the overall operation is not so clear. Why, in a period of economic crisis, in which union membership has begun to fall, should the union promote a vast organisational reform involving the multiplication of structures, and hence an expansion of the organisation, in which it will presumably be difficult to cope with rising costs and falling income? Why dismantle the provincial structures which have always constituted the traditional point of reference for most workers? Why artificially create new sectoral groupings among the federations out of groups, which, in many cases (especially in the CISL and UIL), had very distinct, and not easily reconcilable, identities and traditions? Why try to extend and revitalise peripheral structures in the workplace, especially the factory councils, which provide a lot of opportunity for autonomous initiatives, not easily controlled and co-ordinated by the union, instead of plumping for greater organisational centralisation?

Why Did Organisational Reform Take These Features?

The most widely-held explanations in union circles regarding the significance and peculiarities of the organisational reform can be reduced to two broad types.

From one point of view the reform is seen as the right choice taken at the wrong moment. The project was a design for organisational reform worked out at a time of union expansion, but then implemented during a period of growing, and perhaps unexpected, difficulties. There is no doubt that the union enjoyed an extraordinary expansion at all levels for the entire first part of the 1970s, right up until 1977 (Romagnoli, 1982). It is highly probable that this expansion influenced the reform proposals. Another important and influential circumstance of the early 1970s was the proposed administrative reform of the State, especially of local government. This reform which had aroused a great deal of debate, and was especially welcomed by the unions, resulted in the establishment of regional structures of government in 1970. There were also plans to abolish the province as an administrative unit, replacing it with a smaller unit - the district. This part of the reform was never realised. Nonetheless, the unions thought it opportune to bring their own organisation into line with the projected design of governmental organisation. This point becomes even more relevant if one bears in mind that the unions were interested in reinforcing their role as interlocuter with the state at all levels (in order to play a decision-making role in social policy programmes, labour market issues, etc.).

The above circumstances explain some elements of the union's organisational reform. But why, then, was the reform not modified once it became clear that the unions were not going to continue to enjoy indefinite expansion? Possibly, in 1979, the union was still not aware of the decline in membership, or at least of new members. It is also a possibility that the immediate objective of internal reorganisation seemed far more important than any "minor" preoccupations regarding financial difficulties in the future. On the other hand, it was certainly clear to the unions that the government had set aside any immediate plans it had

had to reform the provincial government. Why, then, did the union still insist on dismantling its own provincial structures, replacing them with districts, which would have no counterpart either in government or within the employers' organisations?

The second type of explanation of the reform, current in union circles, does not necessarily exclude the first. It focuses on the fact that the decision on organisational change taken at Montesilvano constituted a compromise of the different positions of each confederation. There was no disagreement about the need to set up efficient regional structures, but there was much less agreement on the form that sub-regional structures should take. The CGIL was much keener on having small-scale district structures closely linked to the workplace than was the CISL (7). The eventual solution, to have districts slightly smaller than the existing provinces, and to set up, wherever possible, unitary zonal structures, was, as often happens, little more than the sum of the various positions. Any such decision, arrived at in similar circumstances, is likely to be fairly rigid and unamenable to change, even when later external circumstances suggest that a change might be advisable.

Both types of explanations focus on important and relevant aspects of the circumstances surrounding the proposals for reform. But an additional, and somewhat more complex, interpretative hypothesis can also be put forward. As we have seen, the reform was originally based on the changing organisational interests of the unions. These interests (made more urgent by the economic crisis) involved the assertion of greater co-ordination and control over union action. They also followed a long phase in which the prevalent tendency was that of the decentralisation and fragmentation of initiatives. There is evidence of these interests, though it does not always emerge clearly, in union documents. There is also evidence in the "watch-word" of the reform - the "new confederality". Continuing the argument, the specific operative choices taken by the union can be seen not so much as errors of judgement or incongruous outcomes (which is implicit in the two earlier interpretations), but instead as the path intentionally pursued by the union to achieve a better overall co-ordination. In fact, by making certain organisational choices, the union has tried to improve the internal distribution of resources and to exert greater control over the new power centres (at the local level or of specific groups of workers). By this means, the union hoped to eliminate the major points of resistance (which hinder the development of workers' solidarity) to any overall strategy that it might have had.

From this perspective, the breaking-up of the larger provincial structures becomes understandable. Also understandable, is the absorption of small sectoral federations with a disproportionate concentration of power. In the larger groupings, their power will become diluted and constricted by having to take into account the interests of other (less favoured) groups of workers. In both cases, it is a question of finding a counter-balance to, or of neutralising the powers of, the strongest structures. Such a change is of value if one's aim is co-ordination. It continues to be valuable even if in the meantime external circumstances change (such as the government's setting aside of its plans for provincial reform). Such external circumstances may render the reform slightly more or less desirable, but they are not sufficient to deter the unions from their chosen course of action. From the perspective of this argument, the plan to breathe new life into grass-roots structures (factory councils and also zonal councils) and extend their presence, is also desirable. Because these structures are "open" and relatively informal, and also popular with the workforce, their extension and revitalisation seems the best way to actively enlist the support of workers in the union strategy. Other circumstances have also indicated to the unions the importance of maintaining close contacts with the grass-roots. Certainly, a fall in membership as happened in the late 1970s is a warning not to lose contact with the base. So too was the lukewarm welcome given by workers to the new union strategy that was unveiled at the Eur Assembly.

An Interpretation of the Union Strategy

Put another way, the organisational reform proposed by the Italian unions can be seen as a means of achieving the co-ordination of union initiatives without having to turn simply, as one might have expected, to a greater centralisation of industrial relations. The latter would have involved the weakening or virtual suppression of the peripheral levels of the union.

Centralisation or Decentralisation?

One could perhaps discuss interminably whether a centralisation or decentralisation of the system has been proposed by the unions. Many arguments could be used to support either thesis. In practice, what has taken place is a dilution of the organisational system, through the multiplication of internal structures, and the redimensioning and subdivision of the strongest structures.

The union's strategy appears even more reasonable when one takes into account the nature and history of the Italian trade union movement.

As far as regards the nature of the movement, it is important to emphasise that representation and membership is voluntary. The representative capacity of Italian unions is not derived from office, nor is it conferred by legislative recognition. It is also not derived by mandate for extended period of time (periodically subject to verification or election, as occurs in the political representation of citizens in liberal democracies). On the contrary, it is founded in a mandate that can be withdrawn at any time by the workers, who may proceed to try to satisfy their demands by other means (through individual bargaining, direct agreements with their bosses, through the courts, or even organising themselves outside the framework of the union and trying to exert pressure through the parties). It is true that over time the unions have succeeded in obtaining certain safeguards of their right (or legitimacy) to represent the workforce, mainly through official recognition by government and employers. But this right is only valid in the short-term. Over the longer term, the degree of support which the unions are able to obtain from the base determines how much weight their bargaining partners (employers and government) will give to their views, and how much they will involve them in the decision-making process, at least in those decisions which affect their members. If union negotiators are not supported by their members, it is not only possible for employers to negotiate more favourable agreements directly with the workers; but also, in other circumstances, employers may worry that a union will be incapable of making their members accept the results of their negotiations.

The union, therefore, is a representative organisation that continually needs to maintain (and increase) its following among the workforce, or at least it needs to avoid the manifestation of too much internal discord. The position of the unions becomes even more difficult where union pluralism exists, as in Italy. In this case, workers have the possibility not just to leave the union to which they belong, but also to join another. If they are capable of doing so, they may also set up a new union.

This is not the place to discuss the ways in which the confederal union tries to keep the consent of the workforce, while at the same time (and this is the other side of the coin) it has to present itself to employers and government as responsible and able to meet its obligations. However, it is relevant to our argument to point out that the central part of this process is the internal cohesion and co-ordination within the union. Such internal cohesion becomes more important as the union tries to pursue a more general strategy, and one which is less orientated to satisfying the demands of particular groups of workers. Strong internal cohesion enables the union to deal with the difficult task of synthesising the complexity and specificity of grass-roots demands. Such synthesis or reduction depends on a process of successive mediations. In the absence of internal cohesion, each structure will try to maximise its own interests, and attempt to increase its own legitimacy by giving in to grass-roots pressure.

On the other hand, we know (it is the second aspect set out earlier) that the recent history of the Italian union movement is one of decentralisation of the system of industrial relations. This process, as we have seen, resulted in the growing power of the sectoral federations relative to the confederations, and of decentralised apparatuses relative to central ones. How could co-ordination have been increased in order to meet the challenges brought about by the economic crisis without involving all the existing apparatuses? If, in general, the union is an organisation tied to the consent of its members, internally it is also tied to the consent of its apparatuses. There are few possibilities of imposing decisions taken without the consensus of these structures. Thus a simple and clear recentralisation of industrial relations was not a viable alternative. It would probably have aroused the opposition of workers and militants, and the certain opposition of the peripheral apparatuses.

For this reason, the strategy adopted by the unions had to be more complex. It had to increase, at least on paper, and not decrease the functions attributable to the various levels. One may note, for example, the insistence on new objectives for bargaining and territorial action. Though these new functions were set out, the union nonetheless tried to establish a hierarchy of, and functional specialisation between the various levels. This involved, as we have seen, a dilution of the structure, redimensioning the various positions of power previously accumulated, and permitting a reduction of special interests through the process of regrouping. It also involved a general mobilisation of the entire organisation around new themes, which could, at various levels, strengthen the role of horizontal structures and increase co-ordination, in accordance with the principle of "new confederality".

The Unexpected Outcomes: The Resistance From Inside the Union Organisation

Overall the design for reform was very ambitious, but it was also rather risky and its outcome uncertain.

That the project was ambitious, on a grand scale, and did not involve a merely modest internal readjustment, is indicated not so much by the fact that it affected all parts of the organisation or that it brought about massive investments in people and resources, but by the fact that the fundamental objective was to actively involve the entire organisation in the strategy of response to the crisis. Going further, one could even say that the basic idea behind the project was the hope, or ideal (8), that one could create the basis for the involvement of the unions in decisions of economic and social policy, which is necessarily the job of the general secretaries, not merely by imposing decisions from above, but through a process of consensus involving the mobilisation of the entire organisation. Thus the unions could rely on participation to put forward their positions, and not on apathy.

On the other hand, it is clear that the project was also risky and its outcome uncertain. There are two main reasons. First, the organisational change, and the change in strategy which made it necessary, affected the existing power relations and range of responsibilities of various structures. It risked, therefore, arousing the hostility and disaffection of pre-existing sections of the organisation. Second, the entire project had to be directed not by a single and compact union, but by a plurality of unions. Though they were formally unified by the Federative Pact, since the mid-1970s, their positions had begun to be increasingly divergent (Carrieri *et al*., 1983).

Let us consider the first point, which is more immediately relevant. It was precisely to try to avoid hostility and resistance on the part of the apparatus that, in implementing the reform, the union encouraged upward mobility within the union ranks. They also co-opted new cadres for the districts (except where important jobs were involved), and increased the proportion of elected officials (obtained by electoral means in the union congresses of 1981) relative to the numbers appointed by the organisation (Della Rocca, 1983; Regalia, 1982b). This has meant that the majority of union organisers in the provincial structures have moved to the regions (except in a few cases involving the most important districts); that mainly new cadres have moved into the district structures; and that officials and organisers who previously acted without "political" recognition, have benefited from an electoral mandate and moved up into the management group.

In short, the reorganisation of the intermediate structures was carried out in such a way as to provide promotion opportunities for union officials, whilst also giving them political responsibilities. The regrouping of sectoral federations has, in fact, led to a swelling of the ranks of the directing bodies. The price of the regrouping, given that the union did not want to penalise the staff of previously autonomous unions, was the expansion in the size of the structures.

If we try now to set out the practical consequences of the reform, we can see that the first aspect that should be emphasised is the increase in the number of union structures at all levels. According to the Organisational Office of the CGIL, these structures have increased globally from 2,973 to 3,430. The entire personnel (political and technical) of the three organisations in 1982 was over 17,000, approximately one official for every 525 members (9), representing an increase of 20% for the CGIL and 17% for the CISL compared to the staffing levels prior to the reform (Carcano, 1983).

As regards the operational consolidation of the district and regional structures, this appears to have been only partially successful, at least in terms of the prescribed division of functions between the two levels. This point is especially relevant in cases where the traditions of large metropolitan provincial structures were very important (Milan, Rome, Turin, etc.). Only rarely have the newly constituted districts become reference points for the activities of sectoral and zonal organisations. The more successful districts are those which coincide with the metropolitan territories and the provincial capitals. In other cases, they have laboured to take off, and have continued to gravitate in the orbit of the principal cities. Finally, there are few indications so far that the inter-sectoral structures have had much success in tackling, at the local level, the problems of various groups.

As regards the regrouping of the federations, these have been only partially successful; where regrouping has been achieved more rapidly, as in the case of the CGIL, the internal tensions are not insignificant.

The Unexpected Outcomes: The Division Among the Unions

More than the resistance, which was predictable and therefore somewhat discounted, of the pre-existing structures, it was above all the limited internal compactness, the weakness of the federative link between organisations of different strengths and ideological traditions, and the progressive divergence of the positions of each of the major confederations, which posed major problems in the successful implementation of the reform.

In fact, because the reform proved rather costly (at least more costly than predicted) it eventually itself became a source of contention between the major confederations. The UIL, in particular, being the smallest confederation, was unable to come up with the necessary investment in people and means, and fell far short of its commitments. It failed, for instance, to organise congresses of the district structures in 1981, and also failed to establish the districts as properly functioning organisations (Della Rocca, 1983). Often during the congressional debates, the resistance of the UIL to the reforms provided its officials with the opportunity to reaffirm their own historical and organisational traditions. Reciprocal tensions and accusations were also not lacking in the other major confederations.

In the event, in a situation of growing reciprocal diffidence, it was those aspects of the reform which required a joint commitment by all three confederations which became the most difficult to realise. These were the aspects most relevant to the formation and reform of unitary structures. Thus, the factory councils were not extended, as had been proposed, outside those areas in which they were already well established. Likewise, almost nothing was done about the unitary zonal councils, where the commitment to reform had, in fact, been somewhat less firmly expressed. In both cases, the tensions among the unions brought about a situation in which no progress could be made. Worried that they might lose as a result of the changes, the minority organisations were quite content to see the *status quo* preserved.

Thus, an important part of the project, which involved a revitalisation of relations with the grass-roots, and the general mobilisation of the organisation at the peripheral levels, was never realised. This meant that the homogenisation and revision of the rules of operation of the grass-roots structures (namely the factory councils) was also never realised, yet the need for this reform had long been recognised by the union.

We noted earlier the exceptional circumstances in which the factory councils were established. In the mid-1970s, once the phase of rapid growth and collective enthusiasm had passed, and with less possibility of obtaining immediate gains through collective bargaining in the workplace, a union which was now very much stronger than before might have seized the opportunity to make some adjustments to, or rationalise its grass roots structures, in order to bring them into line with its general organisational needs. Such a reform would have implied a better regulation of electoral procedures, a clearer definition of tasks and responsibilities at each level, especially between the union offices inside the workplace and those outside, and the setting-out of precise rules of procedure to adopt in various situations. Most of these objectives were implicit in the reform project launched by the unions in 1979. But the same inter-union tensions which had ruined any possibility of achieving the primary objective of the reform, that is the extension of the factory councils, also prevented any programme from being implemented which might have modified the way the pre-existing structures worked (Regalia, 1983).

It is not true to say, however, that nothing happened. The situation of stalling and non-intervention eventually brought about some quite unexpected and unwelcome results. In the first place, as the favourable conditions in which the councils had been founded changed, they were obliged to make their own adjustments. But they had to do this on an individual basis, without clear terms of reference. They adjusted, then, according to the particular context in which each operated, according to the level of support they could count on among the workers, and also according to the power relations within the workplace and the degree of recognition they received from management. As might be expected, the result was a myriad of different models of industrial relations in the workplace. Such a fragmentation is contradictory with the general requirement of a more efficient co-ordination of union action. In the second place, facing a situation in which the views and positions of the three confederations were increasingly divergent, which itself created a climate of uncertainty, the councils tended to turn in on themselves, closing themselves in their own world of the workplace. Cutting down their ties to the union outside the workplace, they tried to find room for manoeuvre between the pressures of grass-roots demands (to which it is not easy to find a response in time of crisis) and the contrasting solicitations of the outside unions.

Instead of becoming the terminal points of an organisational system able to solicit the active consensus of workers around union decisions, the councils instead ended up by becoming, in the eyes of the unions, nebulous structures whose behaviour could be quite unpredictable, and often disloyal. They became critical points which complicated, rather than facilitated, the process of aggregation of pressures and demands. If, in some circumstances, such as the dramatic dispute at Fiat in the autumn of 1980 (10), the councils have looked like they were losing contacts with the base, in many other circumstances, they have shown, perhaps a bit unexpectedly, that they still continue to enjoy vast support among workers (at least the active ones) and are able to express and interpret their views. One saw evidence of this fact during the general consultations, held by the unions in 1982, to sound out grass-roots opinions on the long-drawn out negotiations with government and employers over the cost of labour. During these consultations, it often transpired that the positions taken by individual factory councils (whether in agreement or disagreement with union positions) obtained the majority of votes in their respective factories. The councils' capacity for mobilisation, even against the wishes of the unions, also emerged dramatically in February-March 1984. This mobilisation followed the clamorous break between the majority Communist wing of the CGIL and the rest of the union movement at the end of the tripartite negotiations to adjust the scala mobile (the mechanism for indexing salaries). When these negotiations broke down, the government issued a decree adjusting the scala mobile. As a result of this very unwelcome government initiative, large numbers of factory councils (and not just those of strict Communist persuasion) assumed the major initiative in organising a nation-wide protest.

A Short Conclusion

Five years after the launch of the most wide-ranging organisational reform ever attempted by the Italian unions, a successful conclusion is even more illusory, especially now that the basic assumptions of the reform are dramatically in crisis. These assumptions concern the unity of purpose of the unitary Federation.

Over the last few years, we have seen recurring manifestations of protest on the periphery every time that the confederal secretaries have been engaged in difficult three-way negotiations with the government over the adoption of anti-inflationary measures. These protests suggest that the results of the reform, to the extent that it has been implemented (or not implemented) are rather like a glove that has been put on the wrong way around.

As we have noted, the basic sense or direction of the entire reform programme can be interpreted as an attempt to revise and dilute the organisation so as to allow, through greater co-ordination at all levels bringing about a "new confederality", a more coherent aggregation of demands. This coherence, however, presupposes a clear division of competences and responsibilities between various levels of the unions, the setting-out of clearly defined procedural rules, and the sanctions that should be applied in instances of transgression of the rules.

Either because of the resistance of pre-existing structures, or even more because of the tensions between organisations, the project of reform was never fully implemented. Above all, no ordering of internal responsibilities was ever achieved. All that really occurred was a multiplication of middle-level structures, involving an expansion of the union organisation. Left intact, and in many senses abandoned, were the peripheral structures.

In the absence of shared procedural rules, the system has become ungovernable. Without such rules, the multiplication of structures has merely given more space for the expression of protest against decisions taken at the centre. Such protest has occurred regularly. It has emanated not just from the factory councils, but also from the district organisations, and the sectoral federations. Against this protest, the control of the centre has shown itself to be completely inadequate.

In conclusion, then, the problem which the unions sought to address when they initiated their reform, that is to succeed in modifying the internal organisation of the union in such a way as to make it more congruent with the requirements of a union movement engaged in the determination of strategies to deal with the country's economic crisis, has still not been resolved.

One should perhaps say instead that the Italian union movement has shown itself once again to be the kind of organisation in which there are lots of opportunities to express protest. There are far fewer possibilities that unwelcome decisions made by the union bosses might be accepted passively and silently.

All this has enormously complicated the decision-making process at the top of the unions, adding to the difficulties of trying to reconcile the divergent positions of the three confederations. As the possibilities of compromise have diminished, the system, as happened in February 1984, has threatened to break apart. We are perhaps at the end of an era in the Italian union movement.

NOTES

1. We shall refer continually in this paper to the large union central offices (CGIL, CISL, and UIL) which organise the vast majority of union members, and which are the privileged interlocutors of government and employers. Besides the unionism of the three major confederations, there are also a number of minor "autonomous" unions in Italy. These smaller unions are mainly to be found in public

employment sectors.

2. The unions established their first regional frameworks when the government established the regions, as a decentralised level of government, in 1970.

3. These were the glorious "camere del lavoro" (according to the CGIL's terminology), the "unioni provinciali", and the "camere sindacali" (according to the terminology of the CISL and UIL respectively).

4. Even today, according to recent research on the factory councils, about 40% of the councils have delegates who are not union members. This applies particularly to delegates representing clerical workers. See Ida Regalia, _Eletti e Abbandonati_, Bologna, forthcoming.

5. These were the regional "frameworks" which were to change from being completely marginal organisations to become centres of power and union initiative at the local level.

6. The province traditionally constituted within the Italian state the most important unit of power at the peripheral level. With the establishment of the regions, its relevance was reduced, at least initially. In addition, new more restricted units of administration were created for the implementation of a number of programmes (in health, education, etc.). These more restricted units consisted of districts, boroughs, etc., and a variety of terms indicating the limited territorial base which they covered.

7. In reality, divergent positions emerged not just between unions, but also within unions. Thus the compromise was much more complex than appears from official evaluations.

8. This at least emerged in some of the more optimistic evaluations of unionists, particularly those in the CGIL (Biagioni, _et al_., 1981).

9. In 1982, union membership in the three major confederations stood at 6,877,106 (counting only actively employed workers). This figure amounts to 46.5% of the workforce (Romagnoli, 1984).

10. In the autumn of 1980, a long and bitter dispute between the unions and Fiat management over the company's plans for internal restructuring ended in defeat for the unions. The union was not helped by discord within the Fiat workforce, which manifested itself most spectacularly in the "march of 40,000", consisting mainly of clerical workers opposed to the union line.

REFERENCES

Biagioni, E. et al., 1981: CGIL anni '80. L'evoluzione delle strutture organizzative, Roma.
Carcano, M., 1983: "L'apparato sindacale: funzioni e compiti" in: Prospettiva sindacale, n. 48.
Carrieri, M. et al., 1984: "La dinamica del negoziato e le relazioni industriali", in: Altieri, G. et al., La vertenza sul costo del lavoro e le relazioni industriali, Milano.
Della Rocca, G., 1983: "Tra continuità e mutamento: la riforma organizzativa del sindacato", in: Prospettiva sindacale, n. 48.
Lange, P. et al., 1982: Unions, Change and Crisis: French and Italian Union Strategy and the Political Economy, 1945-1980, London.
Pizzorno, A., 1978: "Political Exchange and Collective Identity in Industrial Conflict, in: Crouch, C. and Pizzorno, A. (eds.), The Resurgence of Class Conflict in Western Europe Since 1968, London.
Regalia, I., 1978: "Rappresentanza operaia e sindacato: il mutamento di un sistema di relazioni industriali", in: Pizzorno, A. et al., Lotte operaie e sindacato: il ciclo 1968-72 in Italia, Bologna.
Regalia, I., 1982a: "Le rappresentanze sindacali di base", in: Le relazioni sindacali in Italia, Rapporto 1981, Roma.
Regalia, I., 1982b: "Tra rappresentanza e istituzionalizzazione", in: Proposte, CISL Lombardia, n. 3.
Regini, M., 1983: "Le condizioni dello scambio politico", in: Stato e mercato, n. 9.
Romagnoli, G., 1982: "Sindacalizzazione e rappresentanza", in: Le relazioni sindacali in Italia, Rapporto 1981, Roma.
Romagnoli, G., 1984: "Sindacalizzazione e rappresentanza", in: Le relazioni sindacali in Italia, Rapporto 1982/3, Roma.
Salvati, M., 1982: "Strutture politiche ed esiti economici", in: Stato e mercato, n. 4.
Santi, E., 1983: "L'evoluzione delle strutture di categoria: il caso CISL", in: Prospettiva sindacale, n. 48.
Visser, J., 1983: The Unification and Centralisation of the Trade Union Movement: a Comparison of Ten Countries, mimeo.
Streeck, W., 1981: "Qualitative Demands and the Neo-

corporatist Manageability of Industrial Relations", in: British Journal of Industrial Relations, n. 14.

Chapter Eleven

SOCIAL CHANGE AND TRADE UNION MOVEMENT IN THE 1970s

Aris Accornero

The 1970s have rightly been called a decade of "trade union political hegemony" in Italy but this does not hold elsewhere. If we seek to explain this in terms of a historical difference, such as a supposed "backwardness" of the Italian trade unions, then we cannot account for the similarities between Italian and foreign experiences in that decade. Conversely, if we seek an explanation in the political "colour" of Italian trade unionism, we risk magnifying national peculiarities out of all proportion. We would then be forced to attribute all the characteristic features of the Italian union experience to those peculiarities.

Thus another type of explanation is needed. This is what I shall try to offer here, asking first what basic motives underlay collective action by Western trade unions during the decade. An extension of the union movement's tasks has been seen everywhere, in that unions are now engaged in defences of workers' interests on a broader scale and over a longer period of time. Transcending its traditional tasks, the union movement has won recognition of new prerogatives and greater involvement in decision making. Today, in systems of industrial relations everywhere the trade unions are playing more roles with greater stage presence than in the past, as well as counting more heavily in decisions, consensus, and exchange.

So far the Italian case does not seem different. But in Italy this process was accentuated by a more radical strategy which rejected political as well as collective bargaining trade-offs between current and prospective demands. Not only did the union organisations simultaneously demand higher wages and more jobs, but demands for social reforms were superimposed on the bargaining platforms for individual firms. This strategy hinged on a changing technical organisation and social evaluation of labour; and it relied on the social strata most affected by these changes. Operating along these distinctive lines, the union movement won more, and more quickly, in Italy than in other countries, actually encroaching on the process of accumulation and undermining the pattern of social stratification. These very successes encouraged a certain rigidity and a kind of Jacobin-style dogmatism, due to loyalty to the proletarian model in whose name a truly profound and radical transformation was attempted.

In Italy, an attempt was made to institute changes incompatible with the established order - the distribution of income in the country and of power in the factory, prestige and status hierarchies - through a cumulative wave of changes, a democratic revolution led directly by a well defined stratum of workers and a certain type of trade union. This interpretation of the decade thus identifies two factors: the active social group represented, i.e., the mass-worker, with its material interests; and a representative, the working class-orientated union (or "proletarian" union), with its means of expressing that representation. It also depicts the demands put forward as based on a "strategy of reforms" and describes the collective goal as one of "social revolution".

The Mass Worker and Material Interests

First of all, we need to examine how workers' interests were taken up and represented. This calls for an "identification of primary demands which is neither purely pragmatic nor arbitrarily ideological" (Ceri, 1981: p. 72; cf. Lange et al., 1982). Thus we must seek the motivating factors either beyond or before the individual bargaining demands, and beneath the surface of the various contractual "platforms". Likewise the active subjects must be understood as those who actually work to win a given demand, not just those who first conceived it on the basis of their own need.

The job of determining precisely which group raised a certain demand may prove too difficult and relatively fruitless. However, it is feasible, and profitable, to study the type of interest underlying certain claims so that we can attribute them realistically to certain agents, or at least to collective moods. In times marked by powerful collective enthusiasm and dominated by the drive for unity, it is not easy to trace the leading groups behind the interests which emerge; at such times, those who actually raise a proposal or a demand are no more than the spokesmen for everyone. And the union is certainly little inclined to "wonder who these workers are, what they think, what they want - which is actually a way of asking oneself whether the demands have been rightly interpreted" (Ceri, 1981: 73).

The key demands of the early 1970s, which were continued in subsequent phases, were inspired by a powerful desire for equality and security. On the side of equality, we had the abolition of "salary cages" plus equal across-the-board wage increases, followed by the egalitarian changes in the scala mobile, pension mechanisms, and seniority raises, as well as in company bonuses and job classification sytems. On the side of security, a number of guarantees were demanded: control of work pace and the workplace environment; new job categories and better chances for professional advancement; more effective defence of the purchasing power of the lower-income employed and pensioners; job security; equal social status of workers and white-collar employees. The union sublimated egalitarianism, just as it generalised the drive for security, in the satisfying and trusted coupling of wages and power.

The interests underlying these negotiating positions did not consist solely of the unifying, collective demands of the mass-worker, i.e., the ordinary modern production worker. For, in addition to the claims being advanced for oneself, there were also claims against others. In exalting the egalitarian and security aspects of these requests, scholars and trade unionists ultimately neglected a more disturbing implication. This inevitably came to light, however, as the attack on the relative privilege of those just a bit above oneself.

In the unions' negotiating position these motivations were thoroughly sublimated and generalised. Their demands disclosed only the good, constructive aspect, i.e., unity among working people. However, such determined pursuit of wage egalitarianism (sustained over a full decade) involved a truly explosive, socially revolutionary option. And what stood behind this choice was not the usual errors of a "Latin" trade union movement but the interest of a social agent with its own model of society to construct.

The dominant interests represented were those of the mass-worker, symbolising a new social group, an emerging collective identity. Through the union organisation this group drew up a plan "starting from its own particular interest" (Romagnoli, 1981: 46) but also able to draw support from others. This was more or less what occurred in Italy in 1968 and 1969. There was a _statu nascenti_ situation, orientated to new types of political and bargaining demands and involving union mobilisation based on unity and organisation. It is precisely this distinctive genesis which defines the Italian "case": the centre of gravity of the interests represented was formed at a time of powerful collective identification on the part of workers and of students. Such an identification was completely missing in many countries and was anything but easy even in France, much less West Germany, owing to the different actions adopted by the trade unions. Moreover, in Italy this powerful identification lasted for a long time, undiminished in strength.

This worker-student pact, so remote-seeming now, emerged from the moment of rupture of the late 1960s, and was based on the "diffusion among working people of an outlook and an urgent demand for overall social transformation, moving out from the base secured by the new degree of power won in the factory" (Galli, 1980). A rank-and-file with lofty aims and a union looking far ahead, which wanted "to carry worker demands into the heart of society" (Novella, 1974: 44; cf. Accornero, 1974-75) - this was the historic movement that took place in 1969 and that lasted for a full decade. During this time Italian unions drew their "independent capacity to express working class interests" (Regini, 1980: 13, and 1981) entirely from the organisational and political certainty that they represented a homogeneous social grouping, the mass production industrial workers. Thus the trade union movement was able to accept the claims of the rank-and-file and to furnish workers with a collective identity, because it had _chosen_ to represent the most dynamic of the social forces for change - that social group which would soon be known as the mass-worker and conventionally thought of as a metalworker.

The Working Class Union and its Significant Features

Almost from the outset, trade unions sought to

identify themselves with the whole movement for change and to establish their own legitimacy. On the one hand, this effort manifested itself in the factory councils and in the drive for trade union unity, both vigorously endorsed by the metalworkers federations. And, on the other hand, these demands found legal and organisational support in the "Labour Rights Act" (Statuto dei Lavoratori) and in the unions' recently-won privilege of having employers deduct union dues from workers' wage packets.

The "Union of the Councils" and the Labour Rights Act symbolised the identification between the interests and aspirations of the rank-and-file and those of the union as a whole. They stood for a commitment to a new kind of relationship between leadership and rank-and-file. The identity of the unions and the movement for change as a whole and the legitimacy of the unions both supposedly concern everyone - the working class and the working class union - because they are meant to have perfectly identical purposes of social justice. Thus we already see the outline of the proletarian model which would be the reference point for collective action over more or less the whole decade.

For the union, the basis of representation was necessarily assessed in industrial terms: "proletarian unity" was simply the working-class union. Likewise the "Union of the Councils" was seen as the union of the working class interest. If this was the aim of the investiture from the membership, the nature of the relationship with the rank-and-file, and the working class-orientated position of the trade union, there was no presumption of representativeness within the union. But there was more. There was also the very idea of class which was typical of the working-class movement and endorsed and adopted by a majority of the trade union movement at the time.

If the proletariat is an ideological referent which lives in the concrete form of the mass-worker, the metalworker, then the room for action of the working class's organisation, its organisational interest, is no longer restricted. For the role of spokesman for the demands of the entire working class leaves ample scope for union action. In short, the working class-orientated union is the form of organisation which can most easily become independent of the social group it supposedly represents at the same time that it continues to be the best spokesman of its interests. This organisational form consists precisely in the "politicised" union which does not limit itself to the transmission and amplification of demands but also takes the liberty of transforming and even "inventing" them. And this is what the union movement did in Italy for years.

Demands and the Strategy of Reforms

In keeping with a typical orientation of Italian trade unionism, bargaining demands were almost immediately placed in the context of the political demand for new economic horizons. From the very beginning, in fact, the trade union was keenly aware of the limited value of winning any particular claim in the absence of the proper conditions for its consolidation. Thus the slogan, "a new model of development", expressed a strategy which paid attention to the various constraints on the national economy. Similarly, the "new way of making cars" was the protective shell for the initial gains in industrial relations in the factory.

The appeal of these slogans lay in their reference to changes made inevitable by union action and now indispensible to the continuation of that action itself. This concept could not be better expressed than it was by A. Novella, General Secretary of the CGIL, in 1970: "to be substantial and to last over time, union gains must necessarily be extended in the form of reform and power objectives". Further, "to those critics who argue that to ask for reforms and new powers is to engage in politics, we reply that to make only bread-and-butter demands is to endorse mere corporatism" (Novella, 1974: 45-47).

This was a definitive statement of the need to extend the frontiers of, and for, trade union action. Many consequences flowed from this, including political ones, because the approach differed from countries where the labour movement can count on powerful pro-labour parties which have been positively integrated into the system. The new horizons for articulating demands were much broader than could ever have been anticipated or contemplated by the rank-and-file membership on its own. Short-term interests had to be interpreted to link them to long-term trends and orientations and the long-term interest had to be inserted immediately alongside (not in place of) short-term interests. The "proletarian" union has the advantage here, of course, because, although it does indeed represent only a part of the working class, it still retains a leading social role. Thus, at one and the same time, it represents both particular interests and the interests of all workers. For the former, by definition, coincide with the latter. Thus the metalworkers of the industrial North were thought to be the best spokesmen for the interests of the South. Indeed, apart from any sacrifices these workers might make on behalf of the South, the educational and representative role of the union guaranteed this correspondence.

Raising the need to broaden demands in order to defend gains already made, following the simplified, direct route from "factory-to-society", produced a unique kind of representation and negotiation of interests. Once collective action had been initiated, as had already happened as early as 1968 and 1969, culminating in that year's "hot autumn" of struggle, the union reinforced the process by raising its sights and broadening its demands. In practice the substantive claims were intensified due to the need to consolidate economic gains in the wider social framework and there was an expansion of the movement for struggle corresponding to the breadth of the social forces involved. From 1970 the Italian trade union movement pursued this twofold strategy of intensified demands and broadened mobilisation and it also anticipated the interests represented and broadened their aim.

Now, since these claims were stepped up and broadened, the issue arose as to how they could be fitted into the traditional pattern of collective bargaining. The response was a twofold innovation. First, the demands for social reforms (housing, health, taxes) were written into bargaining platforms addressed to the government. And, secondly, the political discussion of these reforms with the authorities was conducted along collective bargaining lines. This institutional innovation, which "unionised" and "contractualised" what had been a typically political field, was underscored by the simultaneous decision that union officials should no longer be members of Parliament.

In short, this was "action external to the collective bargaining market", or political exchange. The definitions adopted in Italy included such terms as "anomalous bargaining", "political confrontations", and "extension of the bargaining model from the factory to the state", though it has been denied that the union "wanted to transpose the collective bargaining outlook mechanically to the level of government institutions".

Does this model really fit the Italy of the 1970s? Descriptively, it would seem so, but in more explanatory terms it seems not. And this reflects an essential feature of the Italian union experience. In Italy, one of the typical conditions for political exchange was apparently lacking - namely, the union organisation's tendency to seek a trade-off between short-term and long-term objectives. The best representation of the interests possible to Italian unions was not to give up present benefits in exchange for future ones, but rather to win rapid changes to consolidate recent working class gains. Nor did this representation consist in the working class renouncing benefits for itself so that there could be redistribution to other social groups: in fact, throughout the entire decade the parable of the two brothers, the employed and the unemployed, was rejected as meretricious and fallacious. It was as if the special interest of the mass-worker could also be satisfied by general gains for everyone. And what is more, this supposedly showed that the interests of working people coincided with the national interest. The working class's demonstration of altruism was the northern metalworker fighting for social reforms, or for jobs in the South. It was not, and must not be, workers' sacrifices.

Underlying this representation of interests, then, there was a <u>proletarian social model</u> managed by the union. The model was so pervasive that it spread from politics to life-style, including an ostentatious shabbiness and coarseness in speech and dress, as if these were the mark of the worker. And was not this same model - not without a certain amount of boastful posturing - reflected in the leftist jargon of the union, with its lack of cultural polish? Was that commonly lofty-sounding language, were those militaristic metaphors, those arrogant proclamations, just the mode of expression of rank-and-file unionism, or of Latin trade union officials? No - for that language also contained the vigorous expression of a desire for transformation that went beyond the form and the traditional <u>modus operandi</u> of even the most militant trade unions.

The general social reference of the model seemed clear, but it was not. It was clear that the union wanted a society with more justice and less privilege, a society based on labour more than capital. But the real aims of the model were not clearly articulated, and at times the union appeared unaware of its purposes.

It was only too clear that the interests of the mass-worker conflicted with those of the bosses. But it was not so clear that these interests necessarily clashed with other, non-antagonistic interests. It is here that we can locate the ambiguity involved in the ideological equation of an abstract working class with a concrete social agent. This identification can work only if the working class interest is adjusted in advance to the national interest; it will fail if it is urged and fought for as the specific interest of the assembly-line worker, who is exploited and has the most to gain from egalitarianism and labour rights. Only out of naiveté, in fact, could one have believed that the unified job classification system and the flat-rate cost-of-living allowance were socially painless because they were just. For every typist who was properly set on the same level as the ordinary worker, however, there was also a better-educated white-collar worker who felt it was unfair to be set on the same level as the skilled worker, and enjoy lesser protection against inflation than the latter. By accentuating the effects of the flattening of wage differentials, Italy's rapid inflation quickly brought the country near the threshold of social tolerance for such a consistently egalitarian policy.

At the same time, working strata outside organised labour also appeared on the scene with their own demands. Such groups as manual workers outside the major centres and big factories, who had not yet been reached by labour organisation, and student workers, who could not be organised in the traditional fashion, brought their own force to reform demands and bargaining claims. The successes of those who had once been the most disadvantaged group in the labour force spurred those currently least orientated to action, starting with the demand for a better or more secure job, or perhaps just for a job. But unions were neither prepared not equipped to organise this less concentrated and less institutional part of the labour force. So there were no forms of association, representation, or defence apart from those already enshrined in working-class tradition. Collective bargaining could only involve regular, full-time jobs, just as initiatives and talks with government authorities were always predicated on the official level of unemployment.

Consequently, even though they made up for past hardships and served as the locomotive for reforms which benefited everyone, the quick gains won by factory workers could also arouse envy in other strata. And if the social group that had formerly been the most disadvantaged could now appear as the most privileged, then, in the long run, this "proletarian" social model could not last. This is the price a working class organisation pays when there is ambivalence concerning working class interests and the national interest, when the working class is supposed to be the "general class". Representing only protelarian interests is a source of strength but can isolate the union, while representing only the national interest - i.e., the "interests" of the economy - generates consensus but can weaken the union.

Thus, as noted, the apparent clarity of the general social reference behind this model was deceptive. Still less clear was the other aspect of the matter: what political model lay behind the social model, symbolised by the ideal-type of the mass-worker and sustained by the calloused hands of the metalworker? Certainly, it was left-oriented, as were the social mobilisation and the electoral advance of those years. But how far left, and aiming at what end? Within what framework of interlocking limits - not just economic but social and political as well - should manual labour, productive labour, be privileged and rewarded, now that virtually everyone seemed to agree in revaluing it? What constraints were to be burst and what constraints imposed? The very demands for equality and security strained not only the enterprise system but also the organisation of society. Moreover, basing a social model on a given working class interest, a given working class composition, clearly implied a certain political alignment, hinging on a certain "social bloc", to use Gramsci's term.

These unsettled aspects of the matter had a number of consequences. Public opinion, for instance, was attracted if not hegemonised by that model, because the assertion of coincidence between working class and national interests, that kind of relationship between worker action and social change - with the union as agent and the metalworker as protagonist - had persuasive charm. For the same reasons, the union leadership, in turn, felt relieved of the need to prove the political compatibility of the social model, which encouraged a certain vagueness in the formulation of claims.

Long lists of reform demands were presented without a single fact or figure. Was this incompetence or bad habit? Can we say that the Italian unions had a culture of protest rather than change? Yes, we can, but aside from the defects of Italy's political culture and political practice, deeply concerned with interconnections and little concerned with quantification, it was the indeterminacy of that sociopolitical model that produced negative effects. This was the reason for the lack of preparation displayed by union officers. And this was also the reason for the utter rejection during those years of co-management - or "conscious bargaining" - with regard to the changes won in the organisation of work. People were operating with reference to a model whose only definite feature was that it had to be "alternative". The end product was a strategy for change totally lacking a model for the transition that was anything more useful than a list of everything the country needed and the demand for the political will to implement it.

Owing in part to the sort of industrial culture that was engendered by the linkage between CGIL and CISL, the Italian trade union confederation was firmly convinced that its gains would stimulate investment, hence employment, and that a socially more just system of production was also necessarily more profitable economically.

Collective Motive and Social Revolution

In the 1970s in Italy, then, conditions were ripe for an unparalleled surge of demands, which is confirmed by the number and size of strikes. What propelled the demands was the widespread agreement by the various strata and social groups which gradually realised the significance of the key demands of the mass-worker, namely equality and security. In the meantime, these profound motives for collective action had been socialised and now constituted the substratum of a more general solidarity, consisting of reciprocal reliance. The goal of union unity spurred mobilisation. And everyone took part in the social struggle movement on the assumption that she/he was equal to the others, in a dynamic, not competitive sense; and on the assumption of guarantees for all against the uncertainties of work and of life. Despite a succession of provocations - first by the right, later by the left as well - this social grouping remained extraordinarily compact, and this was due to the profound cohesive force of those primary motives. It seemed that through its struggles and victories the world of labour was augmenting its internal harmony and its self-confidence in the face of the outside world.

Much more than the mass movements of the 1920s in Sweden or the 1930s in the United States, this movement was of uncertain scope. But to everyone who took part, it seemed the just way, and quite normal for a civilised country. The union's insistence on reforms, in fact, had nothing subversive about it, only a certain Jacobin tinge. And thus the very awareness of what was actually at stake was somehow deadened. Nevertheless, this was a process of democratic transformation so complex and profound as to be comparable to a social revolution.

Two key features marked Italy's experience: the union's strategic anticipation of workers' demands and its substitution of political parties in representing these demands.

Strategic anticipation is what most thoroughly explains the decade-long duration of symbiosis between the material interests of the mass-worker and the modes of expression of the "proletarian" union. First and foremost, anticipation meant the union's getting in harmony with the original demands and with their thrust; expanding them to the substance of demands for social reform; then linking them with one another; and finally, replacing an unmet request or a reform not enacted with a new demand. The result was the kind of mobilisation described above, spurred on still further by the proliferation of bargaining levels and contract expirations.

Given the nature of the claims and the way in which they were channelled, the union had the capacity to anticipate them, that is, place itself concretely ahead of them. This meant that for a long time the union was able to pick up interests just barely emerging anywhere within the working-class alignment and disseminate them quickly among the working masses. It meant, further, that the union was able to propose a number of demands to its membership inspired by credible, shared, collective motives, even though they were rooted in future prospects and not in immediate needs.

This strategic anticipation, however, was not simply upping the bargaining ante. First and foremost, it was a coincidence, a fusion, between the immediate and the deferred components of the demand. A new, united worker-union identity was taking shape, growing from the old roots. And this explains why deferral of demands was inconceivable. This is why there was so little discrepancy between the union's long-term objectives and the immediate interests of the workers. They proceeded together - at least until the so-called "EUR turning point", the 1978 CGIL-CISL-UIL Congress that initiated a strategy of moderation and trade union centralisation.

This pact and this drive engendered that exciting feeling of power that explains the arrogance of union speech in those years. All in all, it was strategic anticipation that made possible a kind of "collectivist" mobilisation around the "proletarian" social model. And it carried a formidable potential for transformation, such as to place constraints on enterprises and on the state with no equivalent compensation on the part of the union. This in turn strongly spotlighted the "discrepancy between what was demanded and what was offered" by working people and by the state and made political exchange extremely costly and virtually impossible (at least until the hopes for trade-off engendered by the emergence of cabinets enjoying Communist support).

Turning to the unions' substitution of political parties in representing demands, this feature has been called specifically Italian. The trade union's substitution for the political system was not due only to the more political nature of the new claims in Italy because it was an international trend, and not a specifically Italian characteristic.

The unions' political role in Italy was not as a substitute for political parties in general; rather, it was as a substitution for the representation performed by the <u>working-class parties</u>. And it was possible precisely because the union federation had become the spokesman for a social stratum and for a concrete working class interest: it had become the means of expression of <u>worker-aims</u>. Hence this substitution involved the temptation to attribute a political motive to working class requests and the tendency to express that motive, based on this active social group - flanking or even replacing the working-class parties.

In fact, that is what happened. The trade union movement took up the representation of the political interests of the mass-worker (and expressed them through the culture of reforms, so congenial to the workers' movement) since the traditional working-class parties did not adequately speak for the social significance of the underlying demands.

Thus with the strategy of reforms, the trade unions not only took the place of other organisations in transmitting an explicit demand but also, owing to their closer identification with the working people revealed a latent demand and anticipated an emerging one which were both lodged in that social group (the mass worker) which put forward rank-and-file interests and sustained their collective identity. This form of trade union representation, which compensated for certain shortcomings in the system of political representation, used a purely trade union weapon, unavailable to the working class parties - the strike - as a means of exerting political pressure. Paradoxically, the use of this weapon politicised not only the claims but also union action itself - striking for reforms is more political than merely demonstrating for them.

The union's presumption of greater legitimacy, partly apparent and partly real - greater than that of the working class parties was in any case validated by the far-reaching mandate it was given, the intense participation of the rank-and-file, and the working masses following it continued to draw in its determined, vigorous action on problems that the working-class parties were also faced with.

In short, what we had was a movement, a kind of insurgency, that represented a genuine democratic transformation and the defence of democracy against the two forms of Italian terrorism, i.e. "fascist plots" and the "armed party" on the left. What it was, then, was a profound but non-traumatic social revolution, shattering in impact but not subversive, furnished with political guarantees as well as political aims.

In 1970, the head of the CGIL described the situation as follows:

> Stimulated by the claims of union action, a profound and radical movement for democratic transformation had grown up in Italy. Until now, the capitalists have imposed the logic, the pace, and the climate that currently characterise the productive apparatus on society as a whole. Now, with the massive wave of struggles, working people want to bring to society a counter-dialectic, namely their capacity for change, their alternative strength. In fact, the factory is becoming once again not

> only an economic engine but a social mainspring, imparting new vigour and quickening the pace of the battle for reforms. This is first effect of the increased bargaining power that trade union movement is achieving with its "return to the factory" (cf. Novella, 1974).

Nor did the union fail to call upon the government "to grasp the dynamism that union battles have impressed upon the economic and social condition of the country in a positive sense". Similarly, there was a very clear recognition that "if there is an institution that is too limited, too narrow for the needs of Italian society, that institution is not the trade union but the state". What was needed, therefore, was a new paradigm of industrial relations, in which "the true factory is the state, the true objective is reform of the state, and the referent of this reform is the political demand" (Novella, 1974).

In good faith, then, the union acted as the holder of a mandate from the working class. So unions assumed the role of a social opposition, trade union unity became the tool of a social bloc for an alternative pattern of development and of society. This took the shape of a tough, determined, even maximalist, but inexpert reformism, heavily conditioned by a certain "workingmenism" on the part of the unions, due in part to the over-extension of the tasks and spheres of representation that followed from the absence of a strong pro-labour party in office.

Conclusion

There was great merit in the tasks the Italian labour unions set for themselves, but there were also great limitations, not only political but also institutional and cultural. I have dwelt on the political limitations, analysing the relations between the union and the power of the cabinet and discussing the political role of the unions and the unions as a political agent. These limitations, comprehensible in part, concerned the unions' resistance to the logic of political compromise and mediation, their refusal to take a position on any given cabinet alignment, sometimes in the name of union independence, sometimes in the name of the unions' political role, which the union was united in asserting. Another typical consequence was the scant attention to "alliances", i.e. the temporary, ad hoc aggregation of related, non-antagonistic interests. This stemmed from the unions' sense of self-sufficiency, engendered by the consciousness of representing a homogeneous, united bloc, fashioned according to the composition of the working class and centred on the mass-worker.

As to the institutional limitations, essentially the political drive inherent in the unions' political action lacked an institutional strategy at all commensurate with the social transformation they sought to pursue. The consequences were all the more serious in that the growing social and union pressure had compressed into a few years all of the political system's internal need for restructuring. Since all that restructuring was able to achieve, however, was a shift in voting patterns (devoid of any appreciable impact on the actual running of the nation except the ambiguous, ambivalent government formula of so called "national solidarity")* in the end the reform impulse carried by worker and social struggles waned - all the more so as that impulse was not welcomed by the PCI, the party that had the greatest benefits from it and was engaged in the difficult experience of a shadow opposition.

Moreover, the unions had never made the social model politically explicit. And perhaps this is why the only political model to appear in these years, the "historic compromise" between PCI and DC, could be apocryphically interpreted as the transcription in party terms of the unity pursued by the unions, whereas in fact it was the utter negation of the central role of the working class in bargaining as in political aspect. So the decade that opened with 1968-69 ended with a disappointed PCI stalking out, angrily slamming the door to a room it had never really entered. That moment, the beginning of 1979, marked the definitive failure of the design for social revolution attempted by the union.

Of course, one cannot fail to note the "excessively generic and sometimes self-contradictory objectives pursued by the union in its 'bargaining' pressure on the institutional political system" and "its tendency to set its claims up as absolutes, and, worse still, on long-term targets". This urgent desire for change embraced not

* From 1977 to 1979, the PCI was part of the Parliamentary majority but was not represented in the cabinet.

only a tumultuous flood of original ideas that did not always work, or the "voluntaristic" raising of new problems when the old ones remained unsolved; this bargaining escalation was also the immediate response to the vitiation of gains on the part of the administrative system after their ratification by the political system. (Critics have observed that Italy's Housing Act "has produced more paper than houses" and concluded that "breaking this barrier implies a radical change in the country's political order and in the direction of the state"). Also, there was the vainly repeated clash with a political system that simply turned a deaf ear, eventually driving CGIL, CISL, and UIL back into the embrace of their own "fraternal" parties.

As for the cultural limitations, mention must obviously be made of the cultural vices and ideological habits that systematically ennobled the collective motivations inspired by the working class and addressed to the country at large, which acquired, by virtue of their ascriptively proletarian roots, a linguistic and logical aggressiveness and "a pathos that enfolds and interprets". Cultural too, were other familiar features of the unions' action over the decade: all-inclusive bargaining, or the extremely high degree of generalisation of targets, and bargaining Titanism, i.e. carrying the logical linkage between one objective and the next to extremes. These features were manifested in the yearning for a universal synthesis, the logical continuum that "welds" together factory and society, collective claims and social reforms.

Another cultural aspect, certainly, was the professional metamorphosis of trade union officers that ensued from the politicisation of the unions. Perhaps it is true, as has been remarked, that too much talk of general problems, economic planning, systemic crisis, investment, got union officers out of the habit of dealing with the more direct issues of collective bargaining. But the culture that sustained this body of trade unionists, so transformed by experience as to have virtually changed functions is not only the natural offspring of the over-extension of the union's tasks, particularly in Italy. It is also the legitimate offspring of the political motivations which the union ascribed to the social subject it represented.

The truly characteristic feature of the Italian trade union in the 1970s was not the corporatist defence of the employed working class but the expression and pursuit of a proletarian social model. We cannot be begrudging on this point. That would mean failing to truly understand the "decade of trade union political hegemony" in Italy. Certainly much disruption derived from it, but so did a lot of positive changes. And could these changes possibly have come without some disruption? And above all without a certain disruption of the established social order? Would Italy really be better off if blue-collar had stayed in their traditional place well below white-collar workers?

The union today, cannot go on representing and protecting only the mass-worker, if for no other reason than that this symbolic figure now seems less decisive, less central than formerly. And this in turn is due, in part, to modifications in the working class structure produced by the union's own action. Nevertheless, unless we bear this social subject, this figure, in mind, we cannot understand the fortunes, or the misfortunes, of the labour union movement in the Italy of the 1970s.

REFERENCES

Accornero, A., 1974-75: Problemi del movimento sindacale in Italia 1943-1973, in: Annali, Fondazione Feltrinelli, XVI.

Ceri, P., 1981: Governabilità democratica e azione sindacale, in: S. Belligni and F. Angeli, eds., Governare la democrazia, Milano.

Galli, P., 1980: Non siamo seduti al tavolo della resa, in: Il Manifesto, May 14.

Manghi, B., 1977: Declinare crescendo, Bologna.

Novella, A., 1974: La rivendicazione operaia nel cuore della società, in: Quaderni di Rassegna sindacale, 51, November-December.

Lange, P., Ross, G., Vannicelli, M., 1982: Unions, Change and Crisis: French and Italian Union Strategy and the Political Economy, 1945-1980, London.

Pizzorno, A., 1980: I soggetti del pluralismo, Bologna.

Regini, M., 1980: Tendenze neocorporative in Italia? in: M. Regini, T. Treu, eds., Sindacato e Stato Capitalistico, Quaderni, 10 (Fondazione Feltrinelli).

Regini, M., 1981: I dilemmi del sindacato, Bologna.

Romagnoli, G., 1981: Ripensare la rappresentanza per riformare l'organizzazione, in: Prospettiva sindacale, 39, March.

Tarantelli, E., 1978: II ruolo economico del

sindacato e il caso italiano, Bari.
Treu, T., 1979: Sindacato e sistema politico, in: Democrazia e diritto, I, January-February.

Chapter Twelve

LABOUR CONFLICTS AND CLASS STRUGGLES

Walther Mueller-Jentsch

Prognoses-Expectations-Illusions

May 1968 in France, September 1969 in West Germany, and Autumn 1969 in Italy opened a new perspective for the leftist (mainly student) protest movement in Western Europe: the working class of the advanced capitalist metropoles did not seem to be as hopelessly saturated and integrated as prominent Marxist theorists like Herbert Marcuse had claimed. The long-lived reconstruction period of post-war capitalism appeared to be over. Class warfare was at the gates; the sharpening of class conflict seemed only to be a question of choosing the proper approach. That the sleeping giant called "the proletariat" was awakening was not contested. The issue was rather the cause: whether it was the falling rate of profit, cyclical economic crises, intra-union rank-and-file and opposition groups, or the leading role of the workers' party which lay behind this awakening.

The first post-war recession in 1966/67 and the political turbulence it caused had already led the West German Left to make some rash predictions. Elmar Altvater stated in <u>neue kritik</u>, the journal of the Socialist German Student Confederation, that the "growth potential of the West German economy was largely exhausted", and he identified the most important "factor limiting growth" as the "qualification structure of labour", an early antithesis to Bravermann. The author had no doubt that "new answers from the working class" were on the agenda (Altvater 1967, p. 25).

At the 16th German Sociologists' Convention in April 1968 in Frankfurt, a group of Frankfurt sociologists impressed those assembled with the following insight, which was also drawn from the

1966/67 recession:

> Even a low unemployment rate of a few per cent leads to voter reactions that can only be controlled with difficulty and to politically dysfunctional consequences. In light of the systemic competition between capitalist and socialist countries, the existence of a reserve army of unemployed would undermine the political system's basis of legitimacy.
>
> (Bergmann et al., 1968, p.74)

The September Strikes in 1969 nourished other dreams. The Sociological Research Institute in Goettingen (SOFI) announced in its study of the September Strikes the prospective "beginning of the reconstruction period of the working class" (Schumann et al., 1971). A research group at Frankfurt based its thesis of the "crisis of co-operative union politics" and its expectations of a change to conflictual, or even class warfare policies on the wildcat strikes of 1969 and 1973 (Bergmann; Jacobi; Mueller-Jentsch, 1975, p. 316 ff.). At about the same time, Prokla (a theoretical journal of the independent Marxist Left) deluded us with the confident prediction that the "material process" (i.e. the consequences for the working class of the accumulation process) was depriving "reformist and integratory beliefs and ideologies of their basis" (Redaktionskollektiv Gewerkschaften, 1972, p. 189).

Like rampaging metastases the concept "class struggle" appeared during these years in the titles and subtitles of books, series and articles: "Sexuality and Class struggle" - "Unions and Class struggle" - "School and Class struggle" "University and Class struggle" - "Social Democracy and Class struggle" - "Labour Law and Class struggle" - "Civil Servants and Class struggle" - "Workers' Education and Class struggle" etc. Labour conflicts were transformed into class struggle and class movements, the analysis of social structure into class analysis, and "class specific" became the most widely used adjective in leftist writings.

The 1970s - The Trade Unions' Decade?*

It was not phantoms alone which fed the hopes and illusions of the Left. Upon what did they base their expectations that a new stage of class conflict had been reached? Principally upon the following phenomena, summarily recapitulated for the entire decade:

- a marked increase in unofficial wildcat strikes at the plant level, especially the two waves of wildcat strikes in 1969 and 1973;
- a significantly higher level of conflict compared to the 1960s: according to the official strike statistics, the number of workers involved in stoppages doubled in the 1970s, while the number of days lost through strikes nearly quadrupled;
- the militant wage policy of the early 1970s, which clearly exceeded the growth and stability-orientated recommendations of the Sachverstaendigenrat (Council of Economic Advisors), and in the end rendered the government's income policy (the "Konzertierte Aktion") ineffectual;
- the increasing conflicts during the triennial works council elections between political factions over the representation of the work force in the works council, which manifested itself mainly in the form of oppositional election slates;
- the transfer of qualitative or non-monetary interests from the plant level to the level of wage negotiations, where they became part of the unions' catalogue of demands as well as being at the heart of several major strikes;
- the counter offensive measures of the employers, called forth by the unions' active wage policy, which were most evident in the increased use of the regional lock-out;
- the continuous rise beginning in the late 1960s (except for the year 1975) not only in union membership, but also in their organisational density, which reversed the stagnant membership trend of the 1960s.

The developments sketched above were the material from which the Left forged its general thesis concerning the "growing intensification of class conflict". Solidly supported by political economy "hardware" - the conditions of accumulation and the falling rate of profit - three interrelated tendencies could be discerned in support of this thesis: 1) the reconstruction of the working class, 2) the end of social partnership, and 3) the crisis

*In this and the following section developments are summarised which are presented in more detail in Mueller-Jentsch 1979 and Brandt et al. 1982.

of the West German model of industrial relations.

The wave of "wildcat" strikes, the politicisation of the works council elections in several large plants, and the rise in the organisational level were only the most spectacular indices of the first tendency, the reconstruction of the working class. Survey results indicated the same thing: one opinion research institute concluded from its polls that there was a "growing willingness for class conflict". Labour researchers and industrial sociologists discovered the so-called "motivation crisis", which soon became a fashionable theme and helped stimulate the debate on the "humanisation of work life". The "dynamic wage consciousness" ascertained by the first Frankfurt union study referred to the higher material claims and solidified expectations of workers. For the authors this was an ambivalent finding. It could be interpreted as an accommodation to conformist ways of thinking; but it also contained an interest potential which, even if initially aimed only at wage militancy, could be politicised for anti-capitalist goals (Bergmann, 1972). On the whole there is much evidence to suggest that a "shift in consciousness" did in fact occur in the early 1970s which led to a more pronounced interest orientation on the part of the workers. This is supported by the fact that hitherto socially marginalised groups emerged at that time out of their secondary existence. The women's movement began to gather strength and influenced the actions and thought of unionised women and working women in general. Even groups of foreign workers appeared for the first time as an active collective, for example during the wildcat strikes of 1973.

The second tendency, the end of social partnership, was in a certain sense the logical consequence of the first. For a workforce conscious of its interests could not easily be bought off with wage offers which were deemed "distributionally neutral" by the Council of Economic Advisers and received the blessing of the "Concerted Action". The intense controversy within the unions over participation in the "Concerted Action" thus seemed logical. The ineffectiveness of this body in influencing wage policy had admittedly become notorious long before the unions, angered by the employers' legal challenge to the 1976 Co-Determination Law, left it in 1977. Owing to the perceived loss of legitimacy in the wake of the wildcat strike waves, as well as the growing oppositional tendencies among their functionaries, several unions felt obliged to switch in the early 1970s to a militant wage policy and regional bargaining. As the number of strikes - both official and unofficial - increased, they lost their novelty status in the Federal Republic. Large-scale wage strikes took place in the chemical and metal-working industries in 1971, among public employees and postal workers as well as in metal-working in 1974, in printing in 1976, and among dockers and metal workers (again) in 1978 (this time in conjunction with qualitative demands). By-products of this switch to wage militancy and regional bargaining were a redistributional success unique in the FRG's history (the share of wages in national income rose by about 10% between 1969 and 1975) and the qualitative collective bargaining policies introduced in several strong regions (prototypically IG Metall's Stuttgart district). During the 1970s five major strikes occurred over non-monetary demands:

- 1973: a nine-day strike in Baden-Wurttemberg's metal industry for additional rest periods for piece workers and other improvements in working conditions (Lohnrahmentarifvertrag II);
- 1978: a three-week conflict (strike and lockout) in the printing industry over the introduction of computerised text systems;
- 1978: a three-week conflict (strike and lockout) in Baden-Wurttemberg's metal industry for protection against dequalification;
- 1978/79: a six-week conflict (strike and lockout) in the steel industry over the demand for a 35 hour work week;
- 1980: a one-week strike by postal employees over work time reduction for shift workers.

With these developments, which certainly disturbed the political and economic elite, the third tendency - the crisis of the West German model of industrial relations - also gained in plausibility. Obviously this "fair weather model" was not up to these new challenges and could no longer guarantee the usual levels of stability and effectiveness. This view was supported by the redistribution brought about by wage bargaining; the transfer of qualitative interests from the plant to the regional and national levels; the general rise in the level of conflict; the violation of legal norms through wildcat and warning strikes; and, finally, the acute representational weakness, i.e., the loss of ability to compromise and command obedience on the part of union hierarchies.

This interpretation of the conflicts of the 1970s has become less and less self-evident as memories of the strikes fade, as economic crisis become the norm, and as it dawns on many that the sporadic resistance of the working class bears no relation to the magnitude of the injuries it has sustained.

Another Interpretation: Self-regulating Accommodation

If, in the second half of the 1970s, under the still powerful influence of the above-mentioned events, the idea of sharpening of class conflict in West Germany seemed plausible, today a different interpretation has gained currency. This recognises that changes during the 1970s in the economic, political, and social-structural arenas confronted the industrial relations system with new problems which called into question the established compromise formulas and negotiation rituals. These shocks did not, however, cause an institutional crisis; instead they activated a process of dynamic accommodation which increased the ability of the system to resolve conflicts.

In order to elaborate this alternative interpretation, it is advisable to explain the criteria and mechanisms on which the stability and effectiveness of the West German industrial relations system is based. In addition to the material rewards which the system distributes, the following structural principles also contribute to its heightened ability to resolve conflicts:

1) The _dual structure of representation_ dating back to the Weimar Republic separates the sectoral collective bargaining system from plant-level interest representation. As a result the conflict between capital and labour occurs in two different "arenas", thus preventing a possible accumulation of conflict potential. In the language of systems theory "functional differentiation" is present here, based on the heterogeneity of interests, actors, and regulations, which facilitates the encapsulation and reduction of conflicts.

2) The extensive _juridification_ of the West German system is also significant. In general, this includes the legal establishment of the dual representation system through the institutions of free collective bargaining (_Tarifautonomie_) and a

works "constitution" (Betriebsverfassung), as well as the detailed legal regulation of labour conflicts and plant-level industrial relations. The last two areas belong to the sensitive zones of conflict between capital and labour: pacifying them through legal means helps to channel and depoliticise unavoidable social confrontations and thus encourages the professionalisation of conflict management.

3) A further characteristic is the representativeness of the organs of collective interest representation (unions and works councils). Their claim, that they make demands in the name of the membership or workforce at the same time as they remain relatively autonomous from those thereby represented, receives backing from official legal sanctions. This establishes important prerequisites for the ability of unions and works councils to enter into compromises and demand compliance.

4) The pronounced centralisation of the collective bargaining system and its (representative) constituent organisations is another important structural characteristic. Unitary trade union confederations like that of the Federal Republic are able, because of the lack of (or more precisely: only minimal) organisational competition, to aggregate the interests of their members at a high level. The employers also tended to organise themselves in central confederations at an early date. Both sides mutually promoted nolens volens the concentration and centralisation of the other side. In this way a convergence in their vertical structures and a symmetry in negotiating levels was achieved. The result was a centralised bargaining system with large bargaining territories. For the unions this means concentrating on "aggregable", i.e. quantitative, interests to the detriment of qualitative as well as plant- and occupation-specific interests. For the employers, industry-wide agreements have the advantage of standardising wage rates and work time in the sense of setting minimums which still leave adequate room for plant level wage and personnel policies.

If the events of the 1970s did cause a crisis of the industrial relations system, this would be visible in a change, transformation, or elimination of those structural principles which lend the system its political stability and economic effectiveness. It would be difficult to show that this in fact happened. Both sides reacted - following the paradigm of "challenge and response" -

to the changed circumstances and the compulsion to resolve differences of interest through conflict with flexible accommodation strategies; they were supported in these efforts by the actions of the legal authorities. The economic crisis which began in 1974 with its continuous high rates of unemployment has undoubtedly made these adjustment processes easier and even necessary. This will be recapitulated in detail through the examination of four partial processes.

1. Regionalisation of Union Bargaining Policy vs. Centralisation of that of the Employers

Some unions (especially the wage trendsetter, IG Metall) reacted to the legitimation crisis caused by the wildcat strikes of the late 1960s and early 1970s by moving towards a regionalisation of their collective bargaining approach. The union districts (which correspond in general to the West German states) thus received greater autonomy, which allowed them to develop new bargaining concepts and initiatives. The employers' associations responded with counter-measures aimed at greater centralisation. Thus the umbrella organisation of the metal working employers reacted to union regionalisation by creating a "core commission" composed of leading members of the umbrella organisation and its larger state confederations. This commission took part in all regional bargaining sessions and could thus impose a uniform line.

In general the 1970s saw an increasing centralisation and co-ordination in the employers' bargaining policies. Since then the Federal Confederation of German Employers Associations (BDA) has played a much more active co-ordinating role than its counterpart, German Trade Union Confederation the (DGB). Whereas the DGB has at best provided information and other auxiliary services which do not impinge on the bargaining autonomy of its constituent unions, the BDA has used various means to prevent its member organisations from signing what it considers to be unacceptable contracts. These means include the Bargaining Policy Commission (Tarifpolitischer Ausschuss), where the experts of the large industrial organisations are represented, and the "taboo catalogue", which stipulates that the approval of the BDA must be sought before contracts can be concluded regarding certain sensitive areas (like, for example, reduction of the work week to below 40 hours or the extension of co-determination).

During the second half of the 1970s, the centralising tendencies within the unions also gained in strength, as the example of IG Metall illustrates most clearly. The IG Metall executive secured its co-ordination of the bargaining process in two ways: first, by laying down tactical and strategic guidelines and setting mandatory minimum and maximum bargaining demands; and second, by declaring, after top level consultation with the employers' organisation (Gesamtmetall), the first wage agreement reached regionally to be a "model" for all subsequent contracts. This practice of negotiating "model" contracts, which has since become the rule, makes the earlier regionalisation of wage bargaining appear to have been only an interlude.

2. Union Militancy vs. Employer Counter-Offensive

The greater readiness to engage in strike action and the strike experience of the unions in the 1970s as compared to the 1960s expressed itself in the early 1970s in the form of militant wage policies and in the late 1970s as combative qualitative bargaining goals. On the one hand, the "new economic militancy" was connected with the markedly higher inflation rate and more demanding attitude ("dynamic wage consciousness") of workers in general. On the other hand, it must be seen in the context of government attempts to tie union wage demands into an incomes policy. This was a typical constellation throughout Western Europe in the early 1970s and subsequently led in many countries to "wage explosions" following rank-and-file rebellions against their leaders' co-operation with incomes policy schemes. This was also how it happened in the Federal Republic. The September Strikes of 1969 were the beginning of the end for the incomes policy institutionalised since 1967 in the "Concerted Action". The wildcat strikes were followed in 1970/71 by a phase of active, sometimes even militant wage bargaining accompanied by long strikes in the chemical and metal working industries, which led to the highest (nominal and real) wage increases since the founding of the Federal Republic. A renewed attempt by the unions to follow the Concerted Action's wage guidelines in 1972/73 resulted - partly due to an unexpected acceleration of inflation - in a new wave of wildcat strikes aimed at cost-of-living bonuses, this time with stronger overtones of criticism directed at the unions. The next wage round in 1974 was again characterised by major strikes (in the public service sector and the metal industry). The bitter 1976 printers' strike took place in the early stages of the new world economic crisis, which in general contributed to a reduction in wage militancy. If in the early 1970s the Concerted Action had progressively lost its influence over wage determination, the need for it disappeared altogether in the second half of the decade.

Under the pressure from continuous mass unemployment and forced rationalisation, the prototypical demands developed during the period of prosperity were replaced by an orientation towards protecting jobs and securing past material gains. Calls for the protection of income, job, qualification, and wage level gave priority to a "qualitative bargaining policy".

As mentioned already, five major strikes took place during the 1970s over non-monetary demands. Their goals, which at first glance appear to have been quite different, can actually be traced back to similar causes and intentions. The strikes were all the result of the forced technical and organisational rationalisation which had been carried out since the end of the 1960s. In the beginning it was clearly for the most part the un- and semi-skilled as well as the marginal labour force that was affected. It was only later, in the wake of the economic crisis, that skilled workers and other members of the employment "core" were threatened (see Kern; Schauer, 1978). It is well known that the latter also comprise the heart of the unions' membership. Even if the strikes in the metal industry (1973, 1978), printing industry (1978), steel industry (1978/79) and the postal service (1980) had defensive goals (i.e. the prevention of deskilling or wage group reorganisation; postponing layoffs; winning compensatory breaks and vacation shifts in the face of the intensification of work, etc.) they still contained elements of a "social control of rationalisation". They also implied a reduction in the authority of employers to carry out rationalisation instead of simply demanding monetary compensation for it, which had been the practice hitherto.

If the employers' associations had been disturbed by the wage policies of the early 1970s, which had called the established distribution of income into question, then they were made even more uneasy by the "qualitative bargaining strategy", which threatened to attack autonomy in production and investment decisions. The employers' associations did in fact initiate a counter-offensive, and in so doing they were able to bolster significantly their organisational and financial resources for limiting union power. Under the central co-ordination of the BDA they made liberal use of the lockout weapon in the conflict-ridden industries (metal, printing, steel); there were five large-scale lockouts compared to only one in the 1960s. With the frequent use of lockouts the employers had come up with a way of limiting the unions' effectiveness by imposing heavy financial burdens upon them. The costs to the union in lockout payments were normally several times those generated by strike pay. Strike and lockout pay cost IG Metall several hundred million marks in the 1970s. Even the relatively small printers union (IG Druck) had to pay almost 50 million DM to its members during two strikes that were accompanied by lockouts, thus driving the organisation to the verge of bankruptcy.

The high level of organisation and the concentrated sanctioning power of West German employers made direct state intervention in labour conflicts, as practiced for years with mixed success in Britain (where employers are much more weakly organised), superfluous. In addition it could compensate for the early failure of a government incomes policy.

3. "New Mobility" - The Primacy of Collective Bargaining

The heavy burdens placed on union coffers by the employers' use of the lockout force the unions into a more flexible and measured application of their fighting resources. Through changes in their by-laws and modification of the arbitration process the unions in the most strike-prone industries (metal and printing) paved the way for the use of legal and controlled work stoppages of limited duration. Among the by-law changes was the approval of limited strikes called without a strike vote, during which no strike pay would be distributed. The modification of arbitration procedures involved on the one hand their simplification (switch from a two to a one step process), and on the other speeding them up (deadlines for initiating arbitration). The arbitration agreement for the metal industry underwent the most far-reaching changes: arbitration is only to occur when both parties agree to it and there are strict deadlines for both the start of arbitration and the length of the statutory "peace period" (Friedenspflicht) which is limited to four weeks after the expiration of a contract. The set deadlines which the agreement prescribes make the concentration of wage negotiations in a single time period advisable, and allow for the precise termination of limited industrial actions. The shortening of the "peace period" allows for a more flexible use of work stoppages. According to the old agreement, these were only officially permissible after arbitration had failed; thus as a rule only confrontations involving large-scale strikes were possible. The new rules permit a measured application of social pressure. For example, during difficult negotiations warning strikes and demonstrative work stoppages can be called and then further escalated in key plants after the expiration of the four week "peace period". The pressure of deadlines and the use of precisely timed work stoppages make a stronger central co-ordination of regional bargaining appear necessary because it raises both the unions' chances of success and the ability of the employers to resist. Even when the flexible use of work stoppages requires decentralised mobilisation this remains embedded in a central strategic conception. After the new arbitration agreement was concluded, the IG Metall executive inaugurated a strategy known as "new mobility" (neue Beweglichkeit) which was characterised by the formula "central co-ordination and decentralised mobilisation".

The simplification and speeding up of the arbitration process strengthened centralising tendencies on both sides. The modified arbitration rules led in the relations between the two bargaining opponents to greater mobility and organisational room to manoeuvre. In other words, the associations on both sides have re-established the primacy of collective bargaining over arbitration in those very areas where conflict has become more likely.

4. Support Through Labour Law

It is inherent in the logic of the Federal Republic's juridified industrial relations system that labour law has also been flexibly adapted to the changed situation and the higher level of conflict. Just as the legislature had recognised the growing importance and expanded range of duties of the works councillor by revising the Works Constitution Act, the Federal Labour Court (BAG) legally grounded the new strike practices through two key decisions which sanctioned warning strikes and lockouts while at the same time making their usage more predictable for each side.

A decision in December 1976 legalised the use by unions of (limited) warning strikes after the expiration of the "peace period" even if the negotiating and arbitration processes had not yet been exhausted. In this way the union strategy of limited conflict, of the graduated and flexible initiation of work stoppages was legally sanctioned. The BAG was motivated by the belief that legalised warning strikes would help extend the period available for "sounding out" the other side about concessions, and thus possibly prevent large scale strikes.

In their decision in June 1980, the BAG granted to the employers the right (contested by the unions) to make use of lockouts, although with restrictions similar to those on strikes. Lockouts are also subject to criteria of fighting parity and proportionality of means according to this ruling. To this end, quantitative norms were laid down for the first time in German labour law. During a partial strike covering less than 25% of a bargaining region, employers can lock out only 25% of the region's workers; if between 25-50% go on strike, only 50% can be locked out, and a total regional lockout cannot be called if more than 50% of its workers are on strike - lockouts of more than 50% are considered "disproportionate".

The BAG has justified its quantitative norms in a purely functional way. Since particular employers are subject to union attack in the place of the entire employers association when strikes are called in selected firms, the employers must have the possibility of ensuring organisational solidarity through the use of (partial) lockouts. The BAG would not agree to the ban on lockouts demanded by the unions since this would upset the "fighting parity" (*Kampfparitaet*) of the two sides. Lockouts raise the unions' willingness to compromise, while their quantitative limitation limits the risks of escalation and cost explosion.

With both decisions on the law of labour conflicts, the BAG laid down the legal framework for a flexible and calculable use of the strike and lockout. In so doing, it proceeded from the premise that the conflict of interest between the two sides might make conflict necessary in principle, but that it should be contained on the lowest possible level.

To summarise the self-regulating processes and adaptions analysed above in a general formula: under the pressure of altered economic and political circumstances and the necessity of settling differences of interest through conflict, a new set of supports were added to the West German industrial relations system without thereby calling any of its basic structural principles into question.

The functional differentiation of conflict regulation has been only partially threatened by the extension of bargaining strategies to include quantitative demands. Ambitious bargaining strategies such as those followed by IG Metall's Stuttgart district or by IG Druck have been cancelled out by counter-mobilisation on the part of the employers. Union representation of workers appeared for a time to be threatened by an organisational legitimation crisis brought about by the September Strikes, but this was papered over by an offensive strategy (militant wage policy, qualitative bargaining demands). The centralisation of collective bargaining was only temporarily threatened. Aside from the fact that the employers associations stuck with centralisation and even strengthened it, the unions returned in the second half of the 1970s to central and co-ordinated bargaining policies. The juridification of conflictual relations, which was occasionally called into question by wildcat strikes and the union challenge to the legitimacy of lockouts, was modified in such a way during the late 1970s that at least some of the new forms of conflict between capital and labour could be regulated by legal means.

In general one can conclude that the manifest conflicts of interest between the two bargaining opponents have placed higher demands on their capacity for compromise than during the period of economic prosperity. As we have shown above, changes in collective bargaining have made allowances for this by raising the capacity of the industrial relations system to accommodate conflict - albeit at the price of more conflict. This statement must, however, be qualified. Under the specific conditions of post-war West Germany, strikes had become the exception. One could even say that for the public at large they signalled a state of emergency. The rise in the conflict level during the 1970s, although modest by international standards, must thus be seen as a normalisation of the relations between capital and labour, which hitherto had been exceptionally peaceful for a dynamic capitalist society like the Federal Republic. According to this interpretation, the inter- and intra-organisational changes carried out by both sides can be seen as self-corrections of the institutional mechanisms for dealing with conflict called forth by a higher existing potential for conflict. The new legal norms are just a ratification of this.

The stabilising effects emanating from the collective bargaining system have their reverse side - they are based in part on the isolation and diversion of potential conflicts. Because the risk of unemployment falls disproportionately on groups which are poorly organised and have trouble fighting back, the core area of collective bargaining is relieved of pressure. A growing percentage of the population who _can_ work is no longer formally employed. The collective bargaining system has become irrelevant for this segment of the labour force. Both the risk of becoming unemployed and of being unemployed for a long time fall principally on the "problem groups" in the labour market: the handicapped, old people, women, teenagers, foreign workers, etc. The unequal distribution of unemployment and the removal of "problem groups" from the labour market have certainly served to blunt the effect of the on-going employment crisis.

Conjectures on the Future of Unions

The question of the future of unions calls for a speculative answer which I will attempt to justify by presenting five theses. These are conjectures about long term trends and expected, not desired, developments.

I. The unions currently find themselves in a precarious situation which has no historical precedent. Criticism of union politics and actions appears at once justified and unjustified. _Justified_ because the unions have been guilty of many sins of omission both in the past and present; unjustified because profound, even epoch making changes in the socio-economic situation have greatly reduced the unions' room for manoeuvre. The vulnerability of union organisations to the effects of a recession or depression is undoubtedly greater than was believed in the 1960s and 1970s. The "dull compulsion of economic relations" (Marx) forced the unions in most capitalist countries - after an extraordinary period of growth - onto the political and economic defensive. With the sharp deterioration in the economic situation the dependency on the business cycle typical of the "classical" union seemed to return. This was even more true of countries where conservative governments tried to master the crisis through neo-liberal economic policies.

Still in all I have no doubts that unions have a future. At the core they are and remain (whatever else they may be) elementary organisations for the protection of workers. Wherever there is alienated wage labour, there will always be - regardless of the social system - trade unions. If they are repressed through terror, economically corrupted, or politically eliminated, illegal, subversive replacements will appear (the most impressive lesson in this respect is offered by Solidarity). Thus the initial question has been answered: unions do have a future, but what kind of future?

II. We are currently experiencing fundamental transformation in the political and social character of unions in the capitalist West. It has been set in motion by the inevitable erosion of the social substrate and the decline in the political substance of the unions. The main features of this process were already recognisable in the 1960s, although partially hidden by the developments of the late 1960s and early 1970s. They involved the transformation of all-encompassing communities aimed at the emancipation of a class into groups representing particular interests. If the socialist labour movement (of which the unions were an integral part) could once claim with justified pathos that they represented the interests of all mankind in addition to those of a persecuted class, today these claims belong to the past.

I would like to support the idea of a transformation of the unions with four arguments:

1) The traditional union membership base, industrial workers, is in decline. There is an on-going secular shift in the organisation of employment from the industrial to the service

sector; sociologists of opposite political camps like Daniel Bell and Alain Touraine are already talking of a post-industrial society. Two developments here are of special significance for the unions. First, the percentage of civil servants and white-collar employees in the total labour force will continue to rise for the foreseeable future. In many developed countries, this percentage already exceeds 50%. In twenty years it may be as high as two-thirds. Civil servants and white-collar employees are harder for unions to organise than blue-collar workers. Second, older and traditionally unionised sectors like mining, steel, engineering, and printing are in decline.

A serious objection to the contention that "white-collarisation" will weaken the unions is that the on-going rationalisation of white-collar work will bring the market situation and working conditions of employees closer to those of manual workers, thus calling forth a similar organisational response from the former. In answer to this, one can say that empty clerical work, which lacks any professional pride, requires only minimal qualification, and produces non-material products, is not suited, as was traditional industrial labour, to producing trade union consciousness and collective behaviour.

2) The proletarian milieu, which in the past encouraged union participation and cohesion, is being progressively dissolved. With the disappearance of the traditional "workers' culture", the unions lose the force of solidaristic-political cohesion. The decomposition of "proletarianism" is due to the following social processes:

- improvements in the standard of living (for example the large increase in home and apartment ownership among workers);
- social mobility: as a result of the growing need for white-collar workers, the filling by foreign workers of the lowest blue-collar jobs, and the social democratic education reforms, many children of workers have been able to advance progressively into non-manual professions;
- changes in living conditions: in place of the "communicatively intense crampedness" (Mooser, 1983, p. 297) typical of worker housing, and the proletarian sphere to which the street, neighbours, the pub, and the club belonged, "privacy" has today become the dominant lifestyle even for blue-collar workers, thanks to larger apartments, more leisure, and socially mixed neighbourhoods;

- different use of free time to the benefit of the family, and consumption of a "mass culture" free of any class-specific character.

3) The lack of politically incisive class experiences has undermined the political substance of the union organisation. It was never the disadvantages resulting from wage labour alone that led workers to see themselves as members of a social class. It was rather the mechanisms of social repression like class voting systems, class justice, and industrial absolutism which helped blue-collar workers gain a collective social and political identity. In many European countries, resistance against fascism and the battles against reaction and restoration after 1945 provided politically decisive collective experiences.

It was the palpable existence of "two nations - the rich and the poor" (Disraeli), i.e. the sharp social dividing line between working class and bourgeoisie, which created communities of belief out of labour organisations. In the future, the new dividing line in the capitalist countries could run - as is already the case in Reagan's U.S. and Thatcher's Britain - between the efficient and productive on the one side and the weak, unemployed, and marginalised on the other. In the wake of this development, unions will be forced more and more into the role of interest organisations of the "productive core".

4) The recent trend towards a relativisation of job and profession represents a challenge to the unions the consequences of which are not yet predictable. The traditional central importance of work is under-going a qualitative and quantitative relativisation thanks to the following factors:

- The significant reduction in the average working day as a result of both higher productivity and union efforts. Also the shortening of the average active working life to between 30 and 40 years due to a longer educational period and earlier retirement coupled with longer life expectancies.

- The part time employment and flexible working hours preferred by many, as well as the emerging forms of electronic home employment reduce the chances of collective work experience and organisation.

- The frequently cited "change in values" among the younger generation indicates a move away from the traditional orientation based on the conventional work ethic. If one takes into consideration the relative stability of the attitudes acquired at an early age, an increasing banalisation of job and profession is to be expected, in so far as they do not provide chances for self-realisation and individual development.

All these trends have in common the relativisation of the work sphere; they weaken the union substrate which rests on wage labour and the dominance of the work role.

III. If the above mentioned tendencies have been interpreted correctly as to their significance for the unions, the question arises as to how we can view the 1970s in such a way as to justify the characterisation "decade of the unions". Based on what is said above, would not one have expected a further decline in the importance and function of unions, as Galbraith (1967) predicted in the mid-1960s?

In my opinion the unions profited in the 1970s from social developments which they did not cause. The break in post-war continuity which occurred in many Western countries in the mid-1970s led to the breakdown of consensus and the end of restraint. It also implied the need for reforms and the modernisation of the economy. In the Federal Republic, it was primarily the potent mixture of the youth and student protest movement, Keynesianism and social democratic reform policies, the women's movement, and citizens' initiatives which led to the break with the social consensus of the Adenauer era and had positive consequences for the unions. Keynesianism and reform policies widened their room for manoeuvre. They also profited both directly and indirectly from the newly emerging social movements. Thus the proportion of young people and women among union members rose disproportionately during those years; through many channels (union educational programs, political discussion groups and grass-roots organisations, active plant-level work by intellectuals, co-operation between pro-union student groups and union locals) an exchange took place, the quantitative dimensions of which are hard to assess, between West Germany's Extra-Parliamentary Opposition (APO) and the trade unions, which in the end helped the latter's organisational efforts.

IV. The unions are in danger of losing in the 1980s all that they gained in the 1970s. They are at the beginning of a new phase of decline and contraction. This is indicated first of all by their significant membership losses. During the last several years, the unions in the four largest EEC countries (FRG, France, Britain, Italy) have suffered substantial membership decreases. More alarming still for the unions' future is the declining proportion of young members, which leads one to fear that the unions will become organisations of older workers. The level of youth membership in the DGB has fallen from 17.5% in 1979 to 15.2% in 1982. Strike activity has also fallen off during the last five years in the four major EEC countries, and even the "strike prone" Italians and British are no exceptions here (Shalev, 1983).

The surprising renaissance of social movements in the Federal Republic (alternative and ecological movement; peace movement; women's movement) raises the question of why the unions have not profited from them as they did in the late 1960s and early 1970s, and why they have not been revitalised by them. The reason that the spark has not caught is to be found, in my opinion, in the completely different historical situation and the specific context of the new social movements. In the late 1960s, the central motives of the protest movement were the critique of capitalism (one of the most popular slogans was "Capitalism leads to fascism! Capitalism must be done away with!") and the critique of American imperialism. The political development of the SPD and the unions showed that the Marxist-orientated social criticism of the APO could co-exist without great intellectual sacrifices with the SPD's reform policies and the active, sometimes militant bargaining policies of the unions. Both sides started from essentially the same model of progress and rationality - the one interpreting it in a gradualist, the other in a revolutionary way.

The new social movements of the 1980s, however, are directed against industrialism and militarism in general, regardless of their social origin. They are critical of technology and growth, and call into question professionalised, segmented industrial labour, as well as the dependence of income on job and profession. For the unions, on the other hand, wage labour and industrialism are constitutive - the unions are their children. The "right to work" that they have always proclaimed refers to wage labour. All unions, whether reformist or radical, have accepted the negative consequences of industrialism - destruction of the environment, elimination of jobs, disruption of the social infrastructure of neighbourhoods - as unavoidable side effects and have played down the problems resulting from them. His trade was and is considered a worker's most important asset, and the job itself took and takes precedence over the character or product of the work involved (whether nuclear power or weapons). The alternative, ecology, and peace movements are all of a fundamentally different opinion in these questions, and it is thus not surprising that they have and will have a tension-laden relationship with the unions.

For these movements, the unions are an integral part of the bureaucratic framework that exists symbiotically with industrialism. Considering the social and political risks the unions would run if they tried to drop out of the supra-class "industrial growth alliance", it is unlikely that they will do so.

The more unlikely an accommodation between traditional and new social movements, the more likely is a widening of the gap separating the relatively privileged from the underprivileged. To claim, however, that the social system will become weaker because of this is an assumption without empirical foundation. We have Japan as a primary example of the contrary; it shows that a society can retain its stability when 30-40% of the workforce enjoy total employment security, while the rest are socially de-classed. The expectation that social conflict will rise with rising polarisation involves transferring the old leftist hope that economic crisis will sharpen class conflict onto a new objective situation.

V. Unions have a future - at least in so far as we have one. It is the future of an interest organisation which workers can no more do without than drivers can do without an auto club. Their traditional economic and political functions are dragged into the centre of gravity of the on-going structural transformation of work. The quantitative and qualitative relativisation of wage labour, the growth in the informal and alternative economies, and finally the increasing fragmentation of the working class (caused by segmented labour markets, unequal unemployment risks, flexible work time, etc.) are rearranging the familiar topography of work. For a substantial number of jobs, the mechanisms of allocation, gratification, qualification, and representation typical of alienated labour have lost their regulative significance.

Whether the unions, which, since the onset of industrialisation have (with good reason) pursued a policy aimed at standardising employment conditions, will be able to develop instruments, strategies and structures which reach beyond the "productive core" and are able to meet the challenge of varied work situations I leave - despite my scepticism - to the future to answer. I consider it unlikely, however, that the creation of a "class subject" with the help of the unions, which the left had expected based on the experiences of the 19th and early 20th century labour, can ever occur. Political hopes which saw the unions as "proletariat class organisations" and "schools of socialism" have undergone an irreversible disenchantment. It is possible that we find ourselves in a situation similar to the one before Marx, when many "real movements" were still seeking their adequate (theoretical) expression.

REFERENCES

Altvater, E. (1967). Perspektiven jenseits des Wirtschaftswunders II, in: neue Kritik, pp. 13-27.

Bergmann, J.G., Brandt, K., Korber, E.T., Mohl, C., Offe (1969). Herrschaft, Klassenverhaltnis und Schichtung, in: T.W. Adorno (ed.), Spatkapitalismus oder Industriegesellschaft, Verhandlungen des 16. Deutschen Soziologentages, Stuttgart, pp. 67-87.

Bergmann, J. (1972). Neues Lohnbewu tsein und Septemberstreiks, in: O. Jacobi, W. Muller-Jentsch, E. Schmidt (eds.), Gewerkschaften und Klassenkampf, Kritisches Jahrbuch 1972, Frankfurt/M. 1972, pp. 171-180.

Bergmann, J., Jacobi, O., Muller-Jentsch, W., (1975). Gewerkschaften in der Bundesrepublik, Frankfurt/M. 3rd Edition, 1979.

Brandt, G., Jacobi, O., Muller-Jentsch, W. (1982). Anpassung an die Krise: Gewerkschaften in den siebziger Jahren, Frankfurt/M.

Galbraith, J.K. (1967). The New Industrial State, London.

Kern, H. und Schauer, H. (1978). Rationalisierung und Besitzstandssicherung in der Metallindustrie, in: Gewerkschaftliche Monatshefte 29, pp. 272-279 u. 482-489.

Mooser, J. (1983). Auflosung der proletarischen Milieus, in: Soziale Welt 34, pp. 270-306.

Muller-Jentsch, W. (1979). Neue Konflikpotentiale und institutionelle Stabilitat, in: J. Matthes (ed.), Sozialer Wandel in Westeuropa, Verhandlungen des 19. Deutschen Soziologentages, Frankfurt/M., pp. 185-205.

Redaktionskollektiv Gerwerkschaften (1972). Intensivierung der Arbeit in der BRD und Gewerkschaften, in: Prokla 5/1972, pp. 125-198.
Schumann, M., Gerlach, F., Gschlossl, G., Milhofer, P. (1971). Am Beispiel der Septemberstreiks. Anfang der Rekonstruktionsperiode der Arbeiterklasse, Frankfurt/M.
Shalev, M.(1983). Strikes and the Crisis: Industrial Conflict and Unemployment in the Western Nations, in: Economic and Industrial Democracy 4, pp. 417-460.

Chapter Thirteen

WORKERS' REACTIONS TO CRISIS

Rainer Zoll and Enno Neumann

It is well known that many workers believe the solution to the unemployment problem is simply to "kick-out" foreign workers. Less well known, however, is that other workers turn the threat posed by unemployment and economic crisis into aggression - aggression against the unemployed and in some cases against themselves. Our comments on this result of crisis refer to a research project on "Worker's Consciousness in Economic Crisis" in which we have participated. The project set out to analyse the worker's perception of crisis and his reaction to crisis. Workers were interviewed from two contrasting types of plant: factories in the shipbuilding and electronic industries adversely affected by the economic crisis and an automobile plant undergoing expansion. Two rounds of interviews were conducted with an interval of 2 years. The results of the first round of interviews (part of which was published in Zott 1981), revealed two different forms of crisis perception.

THE FORMS OF CRISIS PERCEPTION

Reductionist crisis perception: What the workers in this category said about the causes of crisis and unemployment was fundamentally stereotyped. Examples of this kind of argument are: "unemployment is only being dramatised", "the unemployed themselves are to blame", "all we have to do is get rid of the foreign workers". In this way social problems are reduced to standardised patterns of description and explanation. Such reductionism defines and buttresses the social identity of the individual. Thus unemployment is "not" really a problem for the wage-earner; the unemployed are simply too lazy to work, or there are enough "German" jobs for German workers. The wage-earner can certainly find a market for his services "any old time". This self-constructed security is a kind of whistling in the dark, however, since it does not so much hide his underlying insecurity as provide a positive proof of such insecurity. For the individual reductionist perception serves to deny this insecurity so that his identity as a wage-earner is reinforced rather than put into question. The danger is either ignored and thereby minimised directly or it is recognised as a problem but blamed on "scapegoats".

Coping mechanisms of a totally opposite nature characterise the thematic form of perception. Here the individual recognises the grave consequences of unemployment and economic crisis which affect all workers. The financial social and psychological effects of unemployment are addressed emphatically; he recognises the negative effects of sharpening competition between workers and faces up to its underlying conditions in a society in which labour is a mere commodity. This thematic perception puts the identity of the individual to a severe test, for it requires him to endure fear and ambivalence. Thus this form of perception is much less often seen than the reductionist form.

The same workers were interviewed again in 1982, two years after the first round of interviews (for a detailed report cf. Zoll et al., 1984a). Despite the fact that the crisis had worsened substantially in the plants chosen for the research, few changes in fundamental patterns of perception - either reductionist or thematic - were discovered. Yet some significant changes did occur. The full importance of these changes can only be seen when the various types of crisis perception, reaction to crisis, and their consequences for the individual's praxis are analysed in their inner logic, in their interdependence. This inner logic comes out in the hermeneutic interpretation of the individual interviews (Zoll et al., 1984b). The patterns of arguments and actions established in this way are social interpretation patterns, through which the individual processes reality. The social character of these patterns is evident, for out of the 120 interviews conducted in the second round, only seven patterns of perception and interpretation of unemployment and crisis were recognisable.

Social interpretation patterns are neither the arbitrary inventions of the individual nor mere parroting of what is offered through the media.

Instead they result from the practical dealings of the workers (and all other members of society) with social reality - they result from actual life experience (praxis) and the process of socialisation. What one reads, hears and sees in the media naturally plays a part in these interpretation processes. But these media offerings do not so much define as reinforce the latter. Indeed this must be the case since the media offerings themselves are components of social interpretation-patterns and also originate in analogous processes of coping with social reality.

The processing of unemployment and crisis occurs in terms of these social interpretation patterns. True, certain recognisable stereotypes such as the demand that foreign workers be sent home do point to a specific interpretation-model, but these must be located in the context of the respective interview and in terms of the various lines of argument characterising particular interpretation patterns. The interpretation patterns presented here refer only to a certain segment of social reality, namely economic crisis and unemployment. It is therefore not possible to transfer them to complex social normative orientations without further thought.

SOCIAL INTERPRETATION-PATTERNS OF UNEMPLOYMENT AND CRISIS

Out of the seven interpretation-patterns found in the empirical data, five are reductionist, one is more thematic, and one is mixed. The seven interpretation patterns comprise:

1. Economic Mechanisms

This interpretation-pattern explains the causes of crisis and unemployment in terms of economic mechanisms. At the heart of this interpretation-pattern is the image of a complex system whose individual components are related in diverse ways. This is expressed in such catch-phrases as "it's all part of the same thing", or "it all overlaps". If a defect emerges somewhere in this mechanism, it is only logical that it would affect the whole system. Whilst skyrocketing oil prices were named as the main cause of crisis during the first round of interviews, in the second round high interest-rates in the USA were blamed. But competition from Japan and other "booming" countries and the general conditions of the world market were also mentioned. On the side of domestic production, the accelerating process of rationalisation and its job-destroying effects are cited as causes of unemployment.

Some of those interviewed blame overproduction: the time comes when the market has too much of certain goods. This could start with a company which can no longer sell its goods with "knock-on" effects on its suppliers so that its problems produce a whole chain of economic difficulties and thereby lead to more unemployment. Conversely, the theory of underconsumption is also put forward. Sales decrease as the wage-earners' purchasing power decreases, production must be reduced, unemployment rises, and this leads to a further decrease in purchasing power. Events thus get caught in a vicious circle.

In addition to such economic lines of argument, this interpretation pattern is also characterised by the opinion that measures against the crisis are possible. These measures are mostly expected of the government; but the 35-hour-week or increased wages are also named as goals that should be pursued by trade unions. The government itself is expected to provide subsidies for hard-hit branches of the economy, measures of vocational re-training, lowering of the pension age, or a drop in interest rates. In each case these measures correspond to the causes variously identified as having set the economic crisis-circle in motion.

2. Capital and Labour

This interpretation pattern argues that crisis and unemployment are caused by class conflict. Owners of capital and wage-earners stand opposed to one another with conflicting interests. The capitalist controls the "big money" and means of production and against these the workers can set only their labour-power and their numbers. In this argument, crisis and unemployment result from capital's striving for profit. Businesses keep producing more to obtain a higher profit, and they keep lowering their cost of production through rationalisation. These workers believe that the plants rationalise even in times of prosperity as well as during crisis. While growing rationalisation leads directly to layoffs, a state of overproduction is reached if supply has risen too fast relative to demand. Thus this interpretation pattern sees a direct connection between unemployment and profit-hungry business. In general these workers address mainly the production side of the economic process but they also point to the anarchy of the market - individual firms compete with one another and cannot guarantee a crisis-free economic process.

The measures against economic crisis demanded in this interpretation-pattern are defined by the conflict of interests arising from the relationship between wages and profits - the higher the profit of the entrepreneur, the lower the pay of the wage-earner and vice-versa. Whether or not workers' needs are fulfilled and whether unemployment will fall in this zero-sum game becomes a question of the power-relationship between capital and labour. Unless the wage-earner adopts a policy of action, of ensuring that his interests are considered, economic growth favours only the capitalist.

In this interpretation-pattern, decisive action on the part of trade unions in order to fight for the workers' interests is demanded. Even in crisis the unions should not renounce their main fighting resource: striking. There are some differences in this interpretation pattern with regard to the state: workers in this category agree that capital has functionalised the state for its purpose and/or that capital can blackmail the state. The measures demanded against the crisis differ, however, according to how the SPD is assessed as a workers' party. Some suggest that there is an identity of interests between workers, trade unions, and the SPD and therefore demand stronger representation of wage-earner interests at the political level by means of a reform policy carried out by the party. Others reject the SPD, saying that as a "people's party" (Volkspartei), it is too committed to general interests and therefore cannot pursue the genuine interests of the wage-earners. Thus, either by changing the SPD or by forming a new workers' party, these workers want to nationalise big business as a way of overcoming the crisis. They limit this measure to big industry because they see how inefficient are small rationalised plants in the GDR.

3. Fatalism

In this interpretation pattern, economic crisis and unemployment are seen as inevitable consequences of social dynamics against which nothing can be done.

The state as well as the entrepreneur and trade unions are subject to constraints they cannot escape. Specifically, technological progress which makes manpower more and more superfluous, overproduction of goods, and over-capacity in production are all blamed for the worsening economic crisis and unemployment. Accepting their own fatalistic logic such workers regard measures against, for example, unemployment as pointless. Neither the state nor the unions have the means or the competence to intervene in the economic process. True, these workers are not totally against the 35-hour-week or a reduction of the pension age; they simply do not believe such measures will actually create new jobs.

In the most extreme expression of this interpretation pattern, social dynamics leads to catastrophe, leads to war. In a war, "surplus" wealth in the form of excess means of production and consumption goods - an abundance which causes crisis and unemployment - is destroyed. As the most destructive form of "purification" crisis war thus makes a new economic reconstruction period possible, permits a new boom. This form of "solution" to crisis is certainly not favoured by those interviewed, but they do believe war to be the inevitable outcome.

Fatalism, does not simply create an apathy on the part of the workers in their individual praxis (Handlungs orientierung). Any space energy is devoted either to moving up in the profession or to the family, but social activities are rejected.

Nor are any strategies against crisis on the part of trade unions supported here; the weakening of the workers' ability to reproduce themselves is regarded as unavoidable.

4. Politics as the Cause of Crisis

Workers with the political orientation of a "Bourgeois", of an un-political private man, who make up the vast majority of workers, assume as a rule that the social process will continue free of difficulties as long as it is not disturbed by any interventions from inside or outside. Such workers think that crisis can only develop if the state intervenes into society mistakenly or if it fails to make necessary interventions. This criticism of state intervention ranges from the catch-all claim that "it's all because of politics" to a detailed criticism of laws. An important point of criticism concerns finance policy: "There has been a lot of waste with our money for a long time now: credit here, credit there", meaning specifically either financial help to countries of the Third World or military or social spendings. The state "didn't calculate", "overburdened" the economy, threw it off balance. Other points of criticism concern the American policy of high interest-rates and the more general dependence on American politics, the excessive taxation of small factories, and, in many cases, the state regulation of worker immigration.

It follows that the state itself must undertake the task of correcting the mistakes for which it is to blame. Since the state has "exhausted" its money, it must tighten its belt. In particular aid to underdeveloped countries is named; but many workers also demand the reassignment of military spending to civil tasks. Moreover, since the state kicked the economy off balance by taxing too heavily the "small factories" and the "little man", it must now reverse the procedure and tax above all big business or industrial robots.

The policy on foreign workers must also be corrected by corresponding measures. It is important to note that the calling for stricter norms of worker immigration is not substantiated with racist arguments; racist arguments are even rejected. But this does not preclude the possibility that racist motivations underly this argument.

5. Direct Minimisation of Crisis Phenomena

This interpretation pattern has the fewest problems regarding the cause of crisis and unemployment, for the latter are simply denied. The monthly statistics may show millions of unemployed but the overwhelming majority of these people are not "really" unemployed. This result is obtained by an inventive diminution of the published figures on unemployment. Advocates of this view deduct those who took early retirement at age 59; they deduct housewives seeking part-time work and foreign workers. In addition many unemployed do not even want steady work because they can "make ends meet" very well, for example, by doing illicit work. Nor is much credence given to the published numbers of job openings, for the number of jobs actually open is believed to be much higher than that published by the labour offices. This leaves a remainder of perhaps a few hundred thousand unemployed, and "we've always had that".

Economic crisis is also minimised. The workers in this interpretation pattern say that only certain parts of the economy are affected, mainly the construction industry, and that there have always been difficulties in this respect. At first glance one might guess that the prevailing humour is: "There's nothing to worry about". But this interpretation proves to be deceptive. For example, a worker who thinks that there are enough vacancies for all looks in the paper every day to convince himself of this in his personal case. Thereby he tries to calm himself down in the face of an uncertain professional future by denying the general/character of the unemployment problem. Other workers argue that they have never known or even seen anyone who was unemployed. This line of argument is even found among workers who themselves have been unemployed and who even accept that they themselves will never find work again.

Workers with this interpretation pattern reject any dramatisation of crisis and unemployment and refer to contrary personal experience. The latent insecurity continues nevertheless, and it bores in even deeper, so that the problem must be all the more rigidly denied. In elaborating on his belief that students and pensioners should not be counted as unemployed, one worker claimed to know for certain that unemployment statistics include certain groups of people who are not, according to him, "actually" unemployed. But he could not indicate what these groups were.

Suffering from crisis often leads to the dissolution of hitherto rationally controlled argument and its replacement by an aggressive defence mechanism, which then determines world views and normative orientations. In such cases, the threat of crisis is so strong that the barriers which the subject had erected in the form of this fifth type of mechanism collapse; aggressions are set free and are turned on others ("scapegoats") or on the subject himself (submission).

6. Submission

In this interpretation pattern the causes of crisis and unemployment are seen in the high costs of paying the workers. Since the employer is too heavily burdened, either he must live under the threat of bankruptcy or he must rationalise and so kill jobs, which in turn engenders unemployment and crisis. Thus in this interpretation pattern, there is a widespread fear that the employer cannot afford the necessary wage costs to maintain full employment. "Where do they (the employers) get the money they have to pay every month? 1 gross 3200 marks ... That's a lot of money. No one wants to do without, o.k., I'll start. 50 pfennig less per hour for the average worker." To parry the threat of crisis, an all around "climbing down" of the workers is called for. This means accepting pay reductions, as is clear from the above quotation, but it also means doing without striking, since no demands are to be met anyway.

In such attempts to avoid burdening their employer the workers make, not only financial, but other kinds of concessions as well. For example, they go to work sick and also demand this of their colleagues.

> "Heck, I'll go to work with a fever ... I know we are only hurting ourselves when we go to work sick, but that's my attitude and I can't act against it. And if everyone would do that ..., then we'd have all been helped already."

This acting in the employers interests is not just plain submission, for in most cases it is clear that the worker hopes to be in a better position than his colleagues in the competition for jobs. "I have to see to it that I survive".

Thus submission to the employers' interests is supposed to serve the workers' own interests. And here it becomes clear just how deep the existential insecurity caused by crisis and unemployment goes with those in this category. The form they have chosen of realising their interests is a double-edged knife: they impose major concessions on their standard of living and their health - with the vague hope of not totally collapsing into unemployment. For this they will "do anything".

7. Scapegoats

The interpretation pattern "scapegoats" is especially relevant socially for three reasons: (a) one fifth of the interviewed workers are to be counted in this category; (b) it represents a potential for authoritarian developments in society; and (c) almost all other interpretation-patterns can be replaced or dominated by this one, although this is not a necessary development. While in models three, four, and five the crisis phenomena are either directly minimised or are reduced to stereotyped interpretations, the "scapegoats" pattern is characterised by an indirect reduction. This involves transferring the subjective impact of crisis onto others by means of projection of responsibility. Matters get personalised; specific groups of people are made responsible for unemployment and crisis. Some find the causes of crisis in the "indolence" of those seeking work; they see the unemployed as just plain lazy. Others claim what we have is "unemployment imported from Turkey" and thereby attribute the blame to foreign workers.

For the first group the projection of responsibility onto the unemployed rests on a marked consciousness of efficiency leading to the assumption that "who wants to work, finds work". These workers reverse their premise and conclude that the unemployed do not want to work. The logic of this line of argument is that the guilty must be punished. The measures demanded by this group of workers are sharper control over the subsidies for the unemployed, greater obligation to accept other jobs, and, in many cases, the introduction of compulsory labour (Arbeitsdienst). State intervention has a pronounced authoritarian character here. In this interpretation pattern the state fails because it is too tolerant with the unemployed. Consequently the weak state should be replaced by a strong one.

In the variant where foreign workers are the scapegoats, the argument also finds its logical conclusion in the demand for a stronger state. From the very beginning the foreigners are excluded from the community of those who are entitled to jobs. The problem of unemployment becomes a foreigner problem. If there were no foreigners, there would be no unemployment. The state must now respond to the demand "foreigners out" (Auslaender raus!) and enact it.

These two variants of the "scapegoat" pattern are interconnected in two ways. First, they criticise the taking in of people seeking political asylum, who are reproached for their laziness in addition to their being foreigners; second, they criticise German workers for not "displaying a readiness to take over the dirty work" left to foreigners in the past years. So workers supposedly unwilling to work are to blame that the foreign workers were recruited in the first place.

The theme of hyper-alienation is closely connected with the scapegoat-argument. One finds the same fear which in the time before the Third Reich instilled feelings of being a stranger at home, a stranger in one's own country and of being threatened by what is foreign. "You only hear foreign language!" The logic of this interpretation pattern leads to the demand that the guilt of the foreigners be redeemed by punishment. The punishment is to send them home; this simultaneously solves the problem of hyper-alienation in the native country.

INTERPRETATION PATTERN CHANGE AS CRISIS REACTION

These interpretation patterns have a high, albeit relative, stability. They are patterns of processing social reality, patterns which the individual has been internalising for years and which he has acquired through socialisation. The worker will not change them overnight. Nevertheless confrontation of the interpretation patterns with changing reality can lead to problems: the interpretation pattern, in everyday practice, fails in its task of making sense, of giving a meaning to things; it no longer proves capable of processing the changed social reality - which eventually leads to a change of the individual interpretation pattern.

The changes that have been empirically identified reveal three tendencies: (a) the strengthening of the reductionistic form of perception; (b) a growing, and qualitatively changed, expectation of state intervention; and (c) a strong growth of fatalistic estimations of what is possible during crisis. As a rule, such changes follow a logic already contained in the interpretation patterns themselves. Understandably, this logic is closely connected with the interpretation pattern's fundamentally reductionist or thematic character.

It is alarming that all interpretation patterns except for the second can logically develop towards interpretation patterns characterised by aggression. The first and last interpretation patterns prove to be the most unstable. A few of those interviewed who saw the causes of crisis and unemployment as rooted mainly in economic structures demanded action by the state. After such action failed to appear or the wrong measures were taken, these workers then laid the blame for crisis and unemployment on the state. Disappointment at politics' lack of success makes the state the culprit. The already existing expectation of state intervention is intensified, for now the state must correct its past mistakes.

Other workers, who in the first round of interviews had argued that the unemployed were lazy and therefore had only themselves to blame, i.e. had argued in the framework of the last interpretation pattern, now put less weight on this argument. Having become, or expecting to become, victims themselves, they now see fewer chances successfully to parry being affected, and argue in terms of "fatalism".

Further modification tendencies are found <u>within</u> a particular interpretation pattern. As such these changes do not involve a shift from one interpretation pattern to another. They are nonetheless socially relevant and two examples can be given of the transformation of the second interpretation pattern (capital and labour) into a "world view". Under the influence of repeated negative experiences - for example the recurring failure to carry through an active trade union policy in the plant, or the failure of political actions - the drive to change matters, to play an active part in union activities dissolves, and passivity is the result. The early retirement of many elder workers is the result of such a process of change. In addition there is a tendency in the "fatalism" pattern to accept that the outcome of crisis is war. This outcome is not welcomed by those interviewed but it is regarded as inevitable.

In general, a growth of reductionist interpretation patterns can be seen between the first and second rounds of interviews. The relevance of interpretation patterns characterised by the aggressive displacement of fear is only slightly decreased by the "fatalistic turn" because this change does not involve abandoning the scapegoat argumentation; aggressive accusation is integrated in this case into the "fatalism" pattern.

The interpretation patterns involving the aggressive displacement of fear are limited in their scope for development. True, they can change, but in general the authoritarian potential is not diminished by this change. It cannot be excluded that these authoritarian tendencies become openly fascist but this supposes a change in the public climate, since these workers are basically "fellow-travellers" and not "Fuehrers".

This finding is particularly worrying because it is not only the interpretation pattern itself that points in this direction. The fellow-travellers are reinforced in their ideology by the Taylorisation of responsibility, as W. Lempert described it for the work organisation of big plants (Lempert, 1983). The rationality of the thoroughly organised, elementary processes drowns out the irrationality of the whole, which, the more effectively these processes function, becomes all the more inhumane. Nor is this form of instrumental reason limited to the organisation of factory and administration; the mechanism of competition has already dispersed this into every aspect of daily life and divided the individual himself in a Taylorist manner. "The totalitarian attempt to subjugate nature reduces the 'I', the human subject, to a mere instrument of suppression". (Horkheimer, 1968:258.) An important empirical result of our study is that man functions as a "mere instrument of suppression" not only outwardly - against other men and nature - but also inwardly - against himself, his nature, his affective and cognitive needs. The repression that is inflicted on the individuals, they turn against themselves by denying themselves thinking and sensibility through adopting reductionist interpretation patterns. For reduction to stereotyped explanation-formula is the opposite of thinking, of judgement; and aggression is the opposite of sensibility.

Horkheimer had recognised what is missing if these mechanisms are to change: it is

> people who understand that they are themselves the subjects and underlings of their repression. Nothing is done with the image of the "ignorance of the masses", it is itself part of the whole machinery ... The realisation that everything at this precise moment depends on man's making full use of his autonomy could protect culture from the threatening degradation caused by conforming and unreliable friends, or from the annihilation caused by the barbarian within (ibid).

It is no accident that this sounds like philosophy. It reminds one that "fidelity to philosophy means not allowing anxiety to stunt one's capability of thinking" (ibid).

In all cases where a reductionist interpretation pattern prevailed, a hermeneutic interpretation of individual interviews suggested that the base of reductionism was always one or other form of processing fear by repressing it, by repelling it with defense mechanisms. Crisis increases an existential fear which is also present in "normal" times, such that it is repressed, pushed away by many workers. Only a few workers - and they have thematic interpretation patterns - admit their fear to themselves (and others) and try to deal with it openly. The connection between fear, the form of its processing, and "the capability of thinking", as Horkheimer puts it, is unequivocal in this sense. Common sense suggests that "Fear makes you dumb" but it must be pointed out that it is the specific form of processing fear which engenders reductionist interpretation patterns. Changes in social reality can certainly initiate changes in the reductionist interpretation patterns, changes that have the tendency to dissolve reductionism. However, as long as the basic psycho-structures remain connected with mechanisms of defence and repressing fear, the logic of development towards reductionist (even fascistoid) interpretation patterns established in this project will prevail. The crux of the matter is the processing mechanisms of social fear (cf. Zoll, 1982). The form of the processing of fear is first of all an individual one, but it proves to be at the same time extremely social. It is individual because it can be explained in its peculiarities only by the socialisation and life experience of the individual; it is social because its structures are very closely rooted in social relationships and because it also reproduces these relationships. Therefore, changes in the way fear is processed are a matter for the individual himself, where now everything depends on his bringing about the "full use" of his "autonomy". An economic crisis always constitutes at the same time individual existential crises; these are moments of life experience in which such a "full use of the autonomy" is both required and possible. The social character of individual situations of crisis and fear is revealed in the fact that the help (or often the mere presence) of others who understand fear and stand by those who are afraid can make a major contribution to dealing positively with fear. In other words other people can help to dissolve the reductionist interpretation pattern.

This means that the workers' must fight for a revival of solidarity as the fundamental idea of trade unions - above all at the plant level. It has to fight for it, because it is a matter of survival. It can only overcome its crisis if it dissolves the fatal connection between protection and domination - especially as it cannot offer protection against job loss anyway. The concrete utopia of the wage-earners' solidarity in every day life must take the place of the only partly redeemable promise of protection made by the apparatus of trade unions. Such solidarity in daily life carries with it the hope of dissolving reductionist interpretation patterns. In turn this would change the individual himself and all of society.

REFERENCES

Horkheimer, M. 1968: Zur Kritik der instrumentellen Vernunft, in: Kritische Theorie der Gesellschaft, Frankfurt.

Lempert, W. 1983: Taylorisierung der Verantwortung, in: Frankfurter Hefte, Vol. 38, No. 8.

Zoll, R. 1982: Krise und Solidaritaet, in: Gewerkschaftliche Monatshefte, Vol. 33, No. 4.

Zoll, R., ed. 1981: Arbeiterbewusstsein in der Wirtschaftskrise. Erster Bericht: Krisenbetroffenheit und Krisenwahrnehmung, Koeln.

Zoll, R. et al., 1984a: Die Arbeitslosen, die koennte ich alle erschiessen - Arbeiter in der Wirtschaftskrise, Koeln.

Zoll, R. et al., 1984b: Haupsache, ich hab meine Arbeit - Krisenangst und Identitaet von Arbeitern, Frankfurt.

NOTES ON EDITORS AND CONTRIBUTORS

Editors

Dr Otto Jacobi
Institut fuer Sozialforschung, Frankfurt
(Institute for Social Research)
Research Fellow

Dr Bob Jessop
University of Essex
Lecturer

Dr Hans Kastendiek
Institut fuer Sozilaforschung, Frankfurt
Research Fellow
Privatdozent at Free University of Berlin

Prof. Dr Marino Regini
University of Milan
Professor of Political Science

Contributors

Dr Federico Butera
Istituto di Ricerca Intervento sui Sistemi Organizzativi - RSO, Milan
(Institute for Action Research on Organisational Systems)
Chairman and Academic Director

Dr Giuseppe Della Rocca
Istituto di Ricerca Intervento sui Sistemi Organizzatavi - RSO, Milan
Academic Director

Dr Zissis Papadmitriou
Institut fuer Sozialforschung, Frankfurt
Research Fellow

Prof. Dr Gerhard Brandt
University of Frankfurt
Professor of Social Science

Dr Eva Brumlop
Institut fuer Sozialforschung, Frankfurt
Research Fellow

Dr Ulrich Jurgens
Wissenschaftszentrum Berlin,
Internationales Institut fuer Vergleichende Gesellschaftsforschung
(International Institute for Comparative Social Research)
Research Fellow

Dr Harry Scarbrough
Napier College, Edinburgh
Lecturer in Business Studies

Dr Matteo Rollier
Istituto Ricerche Economiche e Sociale - IRES, Rome

Dr Eric Batstone
Nuffield College, Oxford
Senior Research Fellow

Dr Michael Terry
University of Warwick
Industrial Relations Research Unit of the Economic and Social Research Council

Dr Ida Regalia
University of Milan
Research Fellow

Prof. Dr Aris Accornero
University of Rome,
Professor of Sociology
Director of Centro Studi di Politica Economica, Rome
(Institute of Political Economy)

Prof. Dr Walther Mueller-Jentsch
University of Paderborn
Professor of Sociology

Prof. Dr Rainer Zoll
University of Bremen,
Professor of Trade Union Sociology

Dr Enno Neumann
University of Bremen
Research Fellow

TRANSLATORS

Introduction	Isobel Barnden
Butera/Della Rocca	Susan Fisher
Papadimitriou	Hans-Juergen Meinderink
Brandt	Janet Miller-Goerder
Brumlop/Juergens	Hugh Mosley
Rollier	Martin Slater
Heine	Roger Meservey
Regalia	Thomas Ertman
Accornero	Bill Spaulding
Mueller-Jentsch	Thomas Ertman
Zoll/Neumann	Bill Spaulding

INDEX

Austria 137
Automation 13-20, 40-3, 105, 114, 120, 179-80
  see also Computerisation, Technology
Autunno Caldo (Italy's Hot Autumn, 1969) 5, 60, 117, 119, 191, 195, 196, 201, 177, 205, 244, 247
Autonomous Unions 177, 203

British Leyland i, 10, 95-114, 144
Bureaucratisation of trade union organisations 137-57, 150-57

Capital 5, 9, 11, 46
  and corporate strategy 38, 95, 100, 104, 128
  and industrial relations 29, 35, 46-7, 53, 59, 171-4
  and strategy to unions 10, 29, 35-6, 46, 53, 125, 129, 146-9
  see also Management
Cassa Integrazione 117-19, 125-6
Centralisation 141, 196-218, 206, 242
Class Struggle 4, 5, 238-40
Closed Shop 145, 167
Co-Determination 36, 76, 78, 82-3, 149, 182, 241
Collective Bargaining 3, 5, 35, 102, 160-3, 200, 228, 244, 252
  vs. consultation, negotiation 26, 162
  employers' attitudes 75-6, 146, 164-5
  enterprise level 28-9, 35-7, 51, 66, 75-6, 102-3
  trade union bargaining policy 4, 8, 35-6, 45, 50-4, 56-9, 85-9, 119-20, 122-4, 132, 176-8, 181-3, 218, 227, 230, 247
Communist Party 102, 109, 118, 182, 184-6, 191, 213
Compatibility of union organisation and goals 27, 180, 182, 184, 187, 228
Computerisation 22-3, 37-9, 41-6, 79, 124
  See Also Automation
Concerted Economic Steering 188
Confederality 199, 211
Confindustria 185
Consensus, Consent 11, 78, 87, 104, 131
Conservatives 7-12, 253
  in Government 104, 142, 255
Core workforce 47, 55, 179 179, 188, 247, 252, 255 see also Dualism, Labour Market Segmentation
Corporatism 17, 18, 118, 137-141, 142, 158
Crisis 1, 6, 9-11, 38, 105, 114, 120, 179-80, 191, 195-6, 201, 205, 244, 247
  economic 1, 6, 9, 24, 38, 40, 49, 180, 192, 196, 201, 205, 215, 242, 247,
  in Italian unions 177-9
  legitimacy of unions 124-5
  perceptions of, 261-75
  representational 8, 176, 209, 242, 253

Decentralisation in Italian Unions 199-201

De-Skilling 19, 35, 111
Democracy in Unions 142
Dilution of Unions (Italy) 206-8
Donovan Commission 138-40, 161-2, 169
Dual Economy 8, 21, 43-4, 51, 179, 188, 252, 256
Dual Structure of Representation 4, 50-2, 76, 78, 243
Dualism 63, 190

Early Retirement 11, 86-7, 90, 117
Egalitarianism 178-184, 221, 227
Employers´ organisation 244, 245-6, 248
Eur Strategy 182-5, 202, 231

Fascism/Nazism 3, 5, 74, 180, 232, 257
Factory Councils 177, 186-7, 193-4, 200-1, 203-4, 207, 213-4, 223
Federative Pact 209, 233
FIAT 10, 30-1, 111, 116-32, 179, 193, 213
Fordism 47, 50, 54, 61, 64, 121
Foreign Workers 85-6. 241, 252
France 59, 62, 222, 238, 256

Great Britain 3-6, 9-10, 50-70, 95-115 passim, 137-175 passim, 255, 256

Hegemony 180, 183, 219, 228
Humanisation, Quality of Life 37, 44

Incorporation Thesis 137-57
Industrial Democracy 97, 101, 122, 163, 171, 191-2
Industrial Relations 1-3, 6, 25, 75, 78, 196, 208, 213
Information Disclosure 26-8, 29-31, 83-5, 163, 182
Information Technology 16, 25, 39, 63, 129
International Labour Market 55-6, 63
Italy 3, 5, 10, 23-25, 49, 59, 61-2, 65, 75, 116-133 passim, 176-237 passim

Japan 2, 4, 75, 131, 159
Job Controls 168-70
Job Enrichment, 24, 91, 103-4, 176
Joint Consultation 172-4
Joint Management 27, 101-3

Keynesianism 9, 256
Konzertierte Aktion 241, 246-7

Labour Market 15, 17, 50-70, 85, 118, 183
segmentation 21, 47-8, 52-60, 63, 85-6, 127, 189, 247
Labour Party
in government 96, 141
Labour Process 88, 130
Law 3, 4, 9, 11, 21, 76, 84, 142, 171, 223, 240, 242, 250-2

Management Strategies 11, 54, 61, 63, 101, 104, 113-14, 119, 145-149, 164, 170, 185, 245-6
March of the 40,000 117, 131, 178, 217
Market economy vs product economy 39-40
Mass Worker 5, 123, 125, 127, 178, 200, 220-2, 227, 232, 236
Micro-Electronics 6, 15-16, 22-25, 30, 38-44

National Solidarity 116, 123, 182-7
National Interest 228
Neo-Fordism 50, 55-6, 62, 64, 65, 68

Operaism 60

Participation (in technical organisation) 18, 23, 24-6, 28-33, 101-3, 139, 146-9
Parties and Unions 61, 208, 211, 230-1,
Planning, company plan 192
Political Exchange 123, 219, 226, 231
Productivity 24, 97, 99, 104-6, 108-9, 113-114, 132, 140, 143-4, 146, 148, 176, 190-1
Proletarian Model 220-2, 225-6, 230, 238
Public Sector 144

Quality Circles 29, 131, 147)
Rationalisation 38, 50, 53, 79, 82, 85, 89-90, 97, 105, 107, 128, protection agreement 85-7
Representation, Union 176
Robots 22, 30, 41, 79-81, 99, 116-7, 128 178-9, 183-5, 187
Rimini Strategy 181
Ryder Plan, 97-8

<u>Scala</u> <u>Mobile</u> 10, 178, 184-6, 188, 214, 221
Seniority 56, 58, 127
Shop Floor Level 46, 76-8, 83, 100-1, 104, 112-13, 122, 126, 132, 137-41, 161, 165, 193, 200, 240
Shop Stewards 97, 100-1, 108-9, 137-74 <u>passim</u>, 150-9, 161 <u>see</u> <u>also</u> Workers´ Delegates Committees, Works Councils
Skill requirements 5, 35-45, 53-4, 59, 91, 99, 110-11
  and computerisation 37
  deskilling 112
  training 84
Social Contract 138, 142
Social Democracy 9, 37, 178, 186, 257
Social Movements 9, 256-7
Social Pact 187
Social Partnership 5, 92, 241
State and industrial relations 2-6, 11, 108, 142, 171-2, 184, 248
  and economy 6, 11
  and industrial strategy 97, 104, 106, 109, 114, 142, 190, 192, 233
  and unions 9, 49
Strikes 36, 77, 107, 140, 143-4, 229, 240, 242, 246, 248,
Students 222, 238, 242, 256
Sweden 137, 230

Taylorism 2, 19, 24, 41-3, 47, 79, 112-14, 121-2, 128. 130-1
Technology 1-3, 15-17, 22-3, 37-40, 79, 82-84, 95-6, 98-9, 101, 108, 110-12, 126, 131, 194
Terrorism 232
Time, economy of 35, 53,
Trade Unions 5, 177
Tripartism 10, 76, 92, 185-9, 196

Unemployment 6, 7, 9, 45, 58, 80-82, 85, 87-8, 97, 107, 118-20, 140, 143, 146-7, 149, 158-9, 179-80, 239, 245, 252
Unions
  density 7, 76-7, 143, 147, 184, 240
  Italian 178, 198-200
  and party politics 5, 10, 142, 178, 230
  membership 143
  strategy on technical change 45
Union Stewards (VW) 77
Union strategy 120, 196-7
United States of America 49, 56-7, 59, 62, 255, 257
Unofficial disputes 97, 239-40, 242, 246

Volkswagen 1, 11, 73-92

Wage Determination 88-90, 153
Wage Policy 35-6, 78-9, 88-89, 91, 165, 181, 188, 240-1
West Germany 3, 4, 10-11, 22, 50-54, 57-8, 63-7, 55-7, 73-92 <u>passim</u>, 182, 187, o 222, 238-75
White Collarisation 21-2, 24, 77, 193, 227, 254
Workers´ Delegates Committees (Italy) 18, 29, 31-2, 85, 120-2, 126, 130-3, 139, 146-8, 177-8, 181, 189, 193, 199-201, 202-3, 213
Working Conditions 37, 85, 88, 100-1, 103, 109, 119-20, 121, 179, 190
Working Time 11, 22, 35, 45, 50, 80, 87-8, 245, 255
Works Committee (BL) 103-4, 110, 113
Works Constitution Act 36-39, 45, 84, 243-4, 249-50
Works Councils 48, 51, 75-85, 90-1, 160, 199, 240, 242

Young working class 125, 256

Zonal councils 202, 211